THEORY OF ELECTROMAGNETIC WAVE PROPAGATION

CHARLES HERACH PAPAS

PROFESSOR OF ELECTRICAL ENGINEERING
CALIFORNIA INSTITUTE OF TECHNOLOGY

DOVER PUBLICATIONS, INC., NEW YORK

Published in Canada by General Publishing Company, Ltd., 30 Lesmill Road, Don Mills, Toronto, Ontario.
Published in the United Kingdom by Constable and Company, Ltd.

This Dover edition, first published in 1988, is an unabridged and corrected republication of the work first published by the McGraw-Hill Book Company, New York, 1965, in its *Physical and Quantum Electronics Series.* For this Dover edition, the author has written a new preface.

Manufactured in the United States of America
Dover Publications, Inc., 31 East 2nd Street, Mineola, N.Y. 11501

Library of Congress Cataloging-in-Publication Data

Papas, Charles Herach.
Theory of electromagnetic wave propagation / Charles Herach Papas.
p. cm.
Reprint. Originally published: New York : McGraw-Hill, c1965. (McGraw-Hill physical and quantum electronics series) With new pref.
Includes index.
ISBN 0-486-65678-0 (pbk.)
1. Electromagnetic waves. I. Title.
QC661.P29 1988
530.1′41—dc19 88-12291
CIP

To RONOLD WYETH PERCIVAL KING

Gordon McKay Professor of Applied Physics,

Harvard University

Outstanding Scientist, Inspiring Teacher,

and Dear Friend

Preface

This book represents the substance of a course of lectures I gave during the winter of 1964 at the California Institute of Technology. In these lectures I expounded a number of newly important topics in the theory of electromagnetic wave propagation and antennas, with the purpose of presenting a coherent account of the subject in a way that would reveal the inherent simplicity of the basic ideas and would place in evidence their logical development from the Maxwell field equations. So enthusiastically were the lectures received that I was encouraged to put them into book form and thus make them available to a wider audience.

The scope of the book is as follows: Chapter 1 provides the reader with a brief introduction to Maxwell's field equations and those parts of electromagnetic field theory which he will need to understand the rest of the book. Chapter 2 presents the dyadic Green's function and shows how it can be used to compute the radiation from monochromatic sources. In Chapter 3 the problem of radiation emitted by wire antennas and by antenna arrays is treated from the viewpoint of analysis and synthesis. In Chapter 4 two methods of expanding a radiation field in multipoles are given, one based on the Taylor expansion of the Helmholtz integrals and the other on an expansion in spherical waves. Chapter 5 deals with the wave aspects of radio-astronomical antenna theory and explains the Poincaré sphere, the Stokes parameters, coherency matrices, the reception of partially polarized radiation, the two-element radio interferometer, and the correlation coefficients in interferometry. Chapter 6 gives the theory of electromagnetic wave propagation in a plasma medium and describes, with the aid of the dyadic Green's function, the behavior of an antenna immersed in such a medium. Chapter 7 is concerned with the covariance of Maxwell's equations in material media and its application to phenomena such as the Doppler effect and aberration in dispersive media.

The approach of the book is theoretical in the sense that the subject matter is developed step by step from the Maxwell field equations. The advantage of such an approach is that it tends to unify the various topics under the single mantle of electromagnetic theory and serves the didactic purpose of making the contents of the book easy to learn and convenient to teach. The text contains many results that can be found only in the research literature of the Caltech Antenna Laboratory and similar laboratories in the U.S.A., the U.S.S.R., and Europe. Accordingly, the book can be used as a graduate-level textbook or a manual of self-instruction for researchers.

My grateful thanks are due to Professor W. R. Smythe of the California Institute of Technology, Professor Z. A. Kaprielian of the University of Southern California, and Dr. K. S. H. Lee of the California Institute of Technology for their advice, encouragement, and generous help. I also wish to thank Mrs. Ruth Stratton for her unstinting aid in the preparation of the entire typescript.

Charles Herach Papas

Preface to the Dover Edition

Except for the correction of minor errors and misprints, this edition of the book is an unchanged reproduction of the original.

My thanks are due to my graduate students, past and present, for the vigilance they exercised in the compilation of the list of corrections, and to Dover Publications for making the book readily available once again.

Charles Herach Papas

Contents

4 Multipole expansion of the radiated field 81

5 Radio-astronomical antennas 109

6 Electromagnetic waves in a plasma 169

The electromagnetic field 1

In this introductory chapter some basic relations and concepts of the classic electromagnetic field are briefly reviewed for the sake of easy reference and to make clear the significance of the symbols.

1.1 Maxwell's Equations in Simple Media

In the mks, or Giorgi, system of units, which we shall use throughout this book, Maxwell's field equations[1] are

$$\nabla \times \mathbf{E}(\mathbf{r},t) = -\frac{\partial}{\partial t}\mathbf{B}(\mathbf{r},t) \tag{1}$$

$$\nabla \times \mathbf{H}(\mathbf{r},t) = \mathbf{J}(\mathbf{r},t) + \frac{\partial}{\partial t}\mathbf{D}(\mathbf{r},t) \tag{2}$$

$$\nabla \cdot \mathbf{B}(\mathbf{r},t) = 0 \tag{3}$$

$$\nabla \cdot \mathbf{D}(\mathbf{r},t) = \rho(\mathbf{r},t) \tag{4}$$

where $\mathbf{E}(\mathbf{r},t)$ = electric field intensity vector, volts per meter
$\mathbf{H}(\mathbf{r},t)$ = magnetic field intensity vector, amperes per meter

[1] See, for example, J. A. Stratton, "Electromagnetic Theory," chap. 1, McGraw-Hill Book Company, New York, 1941.

$\mathbf{D}(\mathbf{r},t)$ = electric displacement vector, coulombs per meter^2
$\mathbf{B}(\mathbf{r},t)$ = magnetic induction vector, webers per meter^2
$\mathbf{J}(\mathbf{r},t)$ = current-density vector, amperes per meter^2
$\rho(\mathbf{r},t)$ = volume density of charge, coulombs per meter^3
$\mathbf{r}$ = position vector, meters
t = time, seconds

The equation of continuity

$$\nabla \cdot \mathbf{J}(\mathbf{r},t) = -\frac{\partial}{\partial t}\rho(\mathbf{r},t) \tag{5}$$

which expresses the conservation of charge is a corollary of Eq. (4) and the divergence of Eq. (2).

The quantities $\mathbf{E}(\mathbf{r},t)$ and $\mathbf{B}(\mathbf{r},t)$ are defined in a given frame of reference by the density of force $\mathbf{f}(\mathbf{r},t)$ in newtons per meter^3 acting on the charge and current density in accord with the Lorentz force equation

$$\mathbf{f}(\mathbf{r},t) = \rho(\mathbf{r},t)\mathbf{E}(\mathbf{r},t) + \mathbf{J}(\mathbf{r},t) \times \mathbf{B}(\mathbf{r},t) \tag{6}$$

In turn $\mathbf{D}(\mathbf{r},t)$ and $\mathbf{H}(\mathbf{r},t)$ are related respectively to $\mathbf{E}(\mathbf{r},t)$ and $\mathbf{B}(\mathbf{r},t)$ by constitutive parameters which characterize the electromagnetic nature of the material medium involved. For a homogeneous isotropic linear medium, viz., a "simple" medium, the constitutive relations are

$$\mathbf{D}(\mathbf{r},t) = \epsilon\mathbf{E}(\mathbf{r},t) \tag{7}$$

$$\mathbf{H}(\mathbf{r},t) = \frac{1}{\mu}\mathbf{B}(\mathbf{r},t) \tag{8}$$

where the constitutive parameters ϵ in farads per meter and μ in henrys per meter are respectively the dielectric constant and the permeability of the medium.

In simple media, Maxwell's equations reduce to

$$\nabla \times \mathbf{E}(\mathbf{r},t) = -\mu\frac{\partial}{\partial t}\mathbf{H}(\mathbf{r},t) \tag{9}$$

$$\nabla \times \mathbf{H}(\mathbf{r},t) = \mathbf{J}(\mathbf{r},t) + \epsilon\frac{\partial}{\partial t}\mathbf{E}(\mathbf{r},t) \tag{10}$$

$$\nabla \cdot \mathbf{H}(\mathbf{r},t) = 0 \tag{11}$$

$$\nabla \cdot \mathbf{E}(\mathbf{r},t) = \frac{1}{\epsilon}\,\rho(\mathbf{r},t) \tag{12}$$

The curl of Eq. (9) taken simultaneously with Eq. (10) leads to

$$\nabla \times \nabla \times \mathbf{E}(\mathbf{r},t) + \mu\epsilon \frac{\partial^2}{\partial t^2}\mathbf{E}(\mathbf{r},t) = -\mu \frac{\partial}{\partial t}\mathbf{J}(\mathbf{r},t) \tag{13}$$

Alternatively, the curl of Eq. (10) with the aid of Eq. (9) yields

$$\nabla \times \nabla \times \mathbf{H}(\mathbf{r},t) + \mu\epsilon \frac{\partial^2}{\partial t^2}\mathbf{H}(\mathbf{r},t) = \nabla \times \mathbf{J}(\mathbf{r},t) \tag{14}$$

The vector wave equations (13) and (14) serve to determine $\mathbf{E}(\mathbf{r},t)$ and $\mathbf{H}(\mathbf{r},t)$ respectively when the source quantity $\mathbf{J}(\mathbf{r},t)$ is specified and when the field quantities are required to satisfy certain prescribed boundary and radiation conditions. Thus it is seen that in the case of simple media, Maxwell's equations determine the electromagnetic field when the current density $\mathbf{J}(\mathbf{r},t)$ is a given quantity. Moreover, this is true for any linear medium, i.e., any medium for which the relations connecting $\mathbf{B}(\mathbf{r},t)$ to $\mathbf{H}(\mathbf{r},t)$ and $\mathbf{D}(\mathbf{r},t)$ to $\mathbf{E}(\mathbf{r},t)$ are linear, be it anisotropic, inhomogeneous, or both.

To form a complete field theory an additional relation connecting $\mathbf{J}(\mathbf{r},t)$ to the field quantities is necessary. If $\mathbf{J}(\mathbf{r},t)$ is purely an ohmic conduction current in a medium of conductivity σ in mhos per meter, then Ohm's law

$$\mathbf{J}(\mathbf{r},t) = \sigma\mathbf{E}(\mathbf{r},t) \tag{15}$$

applies and provides the necessary relation. On the other hand, if $\mathbf{J}(\mathbf{r},t)$ is purely a convection current density, given by

$$\mathbf{J}(\mathbf{r},t) = \rho(\mathbf{r},t)\mathbf{v}(\mathbf{r},t) \tag{16}$$

where $\mathbf{v}(\mathbf{r},t)$ is the velocity of the charge density in meters per second, the necessary relation is one that connects the velocity with the field. To find such a connection in the case where the convection current is made up of charge carriers in motion (discrete case), we must calculate the total force $\mathbf{F}(\mathbf{r},t)$ acting on a charge carrier by first integrating the force density $\mathbf{f}(\mathbf{r},t)$ throughout the volume occupied by the carrier, i.e.,

$$\mathbf{F}(\mathbf{r},t) = \int \mathbf{f}(\mathbf{r} + \mathbf{r}',t)dV' = q[\mathbf{E}(\mathbf{r},t) + \mathbf{v}(\mathbf{r},t) \times \mathbf{B}(\mathbf{r},t)] \tag{17}$$

where q is the total charge, and then equating this force to the force of inertia in accord with Newton's law of motion

$$\mathbf{F}(\mathbf{r},t) = \frac{d}{dt}[m\mathbf{v}(\mathbf{r},t)] \tag{18}$$

where m is the mass of the charge carrier in kilograms. In the case where the convection current is a charged fluid in motion (continuous case), the force density $\mathbf{f}(\mathbf{r},t)$ is entered directly into the equation of motion of the fluid.

Because Maxwell's equations in simple media form a linear system, no generality is lost by considering the "monochromatic" or "steady" state, in which all quantities are simply periodic in time. Indeed, by Fourier's theorem, any linear field of arbitrary time dependence can be synthesized from a knowledge of the monochromatic field. To reduce the system to the monochromatic state we choose exp $(-i\omega t)$ for the time dependence and adopt the convention

$$C(\mathbf{r},t) = \text{Re}\,\{C_\omega(\mathbf{r})e^{-i\omega t}\} \tag{19}$$

where $C(\mathbf{r},t)$ is any real function of space and time, $C_\omega(\mathbf{r})$ is the concomitant complex function of position (sometimes called a "phasor"), which depends parametrically on the frequency $f(= \omega/2\pi)$ in cycles per second, and Re is shorthand for "real part of." Application of this convention to the quantities entering the field equations (1) through (4) yields the monochromatic form of Maxwell's equations:

$$\nabla \times \mathbf{E}_\omega(\mathbf{r}) = i\omega\mathbf{B}_\omega(\mathbf{r}) \tag{20}$$

$$\nabla \times \mathbf{H}_\omega(\mathbf{r}) = \mathbf{J}_\omega(\mathbf{r}) - i\omega\mathbf{D}_\omega(\mathbf{r}) \tag{21}$$

$$\nabla \cdot \mathbf{B}_\omega(\mathbf{r}) = 0 \tag{22}$$

$$\nabla \cdot \mathbf{D}_\omega(\mathbf{r}) = \rho_\omega(\mathbf{r}) \tag{23}$$

In a similar manner the monochromatic form of the equation of continuity

$$\nabla \cdot \mathbf{J}_\omega(\mathbf{r}) = i\omega\rho_\omega(\mathbf{r}) \tag{24}$$

is derived from Eq. (5).

The divergence of Eq. (20) yields Eq. (22), and the divergence of Eq. (21) in conjunction with Eq. (24) leads to Eq. (23). We infer from this that of the four monochromatic Maxwell equations only the two curl relations are independent. Since there are only two independent vectorial equations, viz., Eqs. (20) and (21), for the determination of the five vectorial quantities $\mathbf{E}_\omega(\mathbf{r})$, $\mathbf{H}_\omega(\mathbf{r})$, $\mathbf{D}_\omega(\mathbf{r})$, $\mathbf{B}_\omega(\mathbf{r})$, and $\mathbf{J}_\omega(\mathbf{r})$, the monochromatic Maxwell equations form an underdetermined system of first-order differential equations. If the system is to be made determinate, linear constitutive relations involving the constitutive parameters must be invoked. One way of doing this is first to assume that in a given medium the linear relations $\mathbf{B}_\omega(\mathbf{r}) = \alpha\mathbf{H}_\omega(\mathbf{r})$, $\mathbf{D}_\omega(\mathbf{r}) = \beta\mathbf{E}_\omega(\mathbf{r})$, and $\mathbf{J}_\omega(\mathbf{r}) = \gamma\mathbf{E}_\omega(\mathbf{r})$ are valid, then to note that with this assumption the system is determinate and possesses solutions involving the unknown constants α, β, and γ, and finally to choose the values of these constants so that the mathematical solutions agree with the observations of experiment. These appropriately chosen values are said to be the monochromatic permeability μ_ω, dielectric constant ϵ_ω, and conductivity σ_ω of the medium. Another way of defining the constitutive parameters is to resort to the microscopic point of view, according to which the entire system consists of free and bound charges interacting with the two vector fields $\mathbf{E}_\omega(\mathbf{r})$ and $\mathbf{B}_\omega(\mathbf{r})$ only. For simple media the constitutive relations are

$$\mathbf{B}_\omega(\mathbf{r}) = \mu_\omega\mathbf{H}_\omega(\mathbf{r}) \tag{25}$$

$$\mathbf{D}_\omega(\mathbf{r}) = \epsilon_\omega\mathbf{E}_\omega(\mathbf{r}) \tag{26}$$

$$\mathbf{J}_\omega(\mathbf{r}) = \sigma_\omega\mathbf{E}_\omega(\mathbf{r}) \tag{27}$$

In media showing microscopic inertial or relaxation effects, one or more of these parameters may be complex frequency-dependent quantities.

For the sake of notational simplicity, in most of what follows we shall drop the subscript ω and omit the argument $\mathbf{r}$ in the monochromatic case, and we shall suppress the argument $\mathbf{r}$ in the time-dependent case. For example, $\mathbf{E}(t)$ will mean $\mathbf{E}(\mathbf{r},t)$ and $\mathbf{E}$ will mean $\mathbf{E}_\omega(\mathbf{r})$. Accordingly, the monochromatic form of Maxwell's equations in simple media is

$$\nabla \times \mathbf{E} = i\omega\mu\mathbf{H} \tag{28}$$

$$\nabla \times \mathbf{H} = \mathbf{J} - i\omega\epsilon\mathbf{E} \tag{29}$$

$$\nabla \cdot \mathbf{H} = 0 \tag{30}$$

$$\nabla \cdot \mathbf{E} = \frac{1}{\epsilon}\rho \tag{31}$$

1.2 Duality

In a region free of current ($\mathbf{J} = 0$), Maxwell's equations possess a certain duality in $\mathbf{E}$ and $\mathbf{H}$. By this we mean that if two new vectors $\mathbf{E}'$ and $\mathbf{H}'$ are defined by

$$\mathbf{E}' = \pm\sqrt{\frac{\mu}{\epsilon}}\,\mathbf{H} \qquad \text{and} \qquad \mathbf{H}' = \mp\sqrt{\frac{\epsilon}{\mu}}\,\mathbf{E} \tag{32}$$

then as a consequence of Maxwell's equations (source-free)

$$\begin{aligned} &\nabla \times \mathbf{H} = -i\omega\epsilon\mathbf{E} \qquad \nabla \times \mathbf{E} = i\omega\mu\mathbf{H} \\ &\nabla \cdot \mathbf{H} = 0 \qquad \nabla \cdot \mathbf{E} = 0 \end{aligned} \tag{33}$$

it follows that $\mathbf{E}'$ and $\mathbf{H}'$ likewise satisfy Maxwell's equations (source-free)

$$\begin{aligned} &\nabla \times \mathbf{E}' = i\omega\mu\mathbf{H}' \qquad \nabla \times \mathbf{H}' = -i\omega\epsilon\mathbf{E}' \\ &\nabla \cdot \mathbf{E}' = 0 \qquad \nabla \cdot \mathbf{H}' = 0 \end{aligned} \tag{34}$$

and thereby constitute an electromagnetic field $\mathbf{E}'$, $\mathbf{H}'$ which is the "dual" of the original field.

This duality can be extended to regions containing current by employing the mathematical artifice of magnetic charge and magnetic current.[1] In such regions Maxwell's equations are

$$\begin{aligned}
&\nabla \times \mathbf{H} = \mathbf{J} - i\omega\epsilon\mathbf{E} \qquad \nabla \times \mathbf{E} = i\omega\mu\mathbf{H} \\
&\nabla \cdot \mathbf{H} = 0 \qquad \nabla \cdot \mathbf{E} = \frac{1}{\epsilon}\rho
\end{aligned} \tag{35}$$

and under the transformation (32) they become

$$\begin{aligned}
&\nabla \times \mathbf{E}' = \pm\sqrt{\frac{\mu}{\epsilon}}\,\mathbf{J} + i\omega\mu\mathbf{H}' \qquad \nabla \times \mathbf{H}' = -i\omega\epsilon\mathbf{E}' \\
&\nabla \cdot \mathbf{E}' = 0 \qquad \nabla \cdot \mathbf{H}' = \pm\frac{1}{\mu}\sqrt{\frac{\mu}{\epsilon}}\,\rho
\end{aligned} \tag{36}$$

Formally these relations are Maxwell's equations for an electromagnetic field $\mathbf{E}'$, $\mathbf{H}'$ produced by the "magnetic current" $\mp\sqrt{\mu/\epsilon}\,\mathbf{J}$ and the "magnetic charge" $\pm\sqrt{\mu/\epsilon}\,\rho$. These considerations suggest that complete duality is achieved by generalizing Maxwell's equations as follows:

$$\begin{aligned}
&\nabla \times \mathbf{H} = \mathbf{J} - i\omega\epsilon\mathbf{E} \qquad \nabla \times \mathbf{E} = -\mathbf{J}_m + i\omega\mu\mathbf{H} \\
&\nabla \cdot \mathbf{H} = \frac{1}{\mu}\rho_m \qquad \nabla \cdot \mathbf{E} = \frac{1}{\epsilon}\rho
\end{aligned} \tag{37}$$

where $\mathbf{J}_m$ and ρ_m are the magnetic current and charge densities. Indeed, under the duality transformation

$$\begin{aligned}
&\mathbf{E}' = \pm\sqrt{\frac{\mu}{\epsilon}}\,\mathbf{H} \qquad \mathbf{H}' = \mp\sqrt{\frac{\epsilon}{\mu}}\,\mathbf{E} \qquad \mathbf{J}' = \pm\sqrt{\frac{\epsilon}{\mu}}\,\mathbf{J}_m \\
&\mathbf{J}'_m = \mp\sqrt{\frac{\mu}{\epsilon}}\,\mathbf{J} \qquad \rho' = \pm\sqrt{\frac{\epsilon}{\mu}}\,\rho_m \qquad \rho'_m = \mp\sqrt{\frac{\mu}{\epsilon}}\,\rho
\end{aligned} \tag{38}$$

$$\begin{aligned}
&\nabla \times \mathbf{E}' = -\mathbf{J}'_m + i\omega\mu\mathbf{H}' \qquad \nabla \times \mathbf{H}' = \mathbf{J}' - i\omega\epsilon\mathbf{E}' \\
&\nabla \cdot \mathbf{E}' = \frac{1}{\epsilon}\rho' \qquad \nabla \cdot \mathbf{H}' = \frac{1}{\mu}\rho'_m
\end{aligned} \tag{39}$$

[1] See, for example, S. A. Schelkunoff, "Electromagnetic Waves," chap. 4, D. Van Nostrand Company, Inc., Princeton, N.J., 1943.

Thus to every electromagnetic field **E**, **H** produced by electric current **J** there is a dual field **H**′, **E**′ produced by a fictive magnetic current $\mathbf{J}'_m$.

1.3 Boundary Conditions

The electromagnetic field at a point on one side of a smooth interface between two simple media, 1 and 2, is related to the field at the neighboring point on the opposite side of the interface by boundary conditions which are direct consequences of Maxwell's equations.

We denote by **n** a unit vector which is normal to the interface and directed from medium 1 into medium 2, and we distinguish quantities in medium 1 from those in medium 2 by labeling them with the subscripts 1 and 2 respectively. From an application of Gauss' divergence theorem to Maxwell's divergence equations, $\nabla \cdot \mathbf{B} = \rho_m$ and $\nabla \cdot \mathbf{D} = \rho$, it follows that the normal components of **B** and **D** are respectively discontinuous by an amount equal to the magnetic surface-charge density η_m and the electric surface-charge density η in coulombs per meter²:

$$\mathbf{n} \cdot (\mathbf{B}_2 - \mathbf{B}_1) = \eta_m \qquad \mathbf{n} \cdot (\mathbf{D}_2 - \mathbf{D}_1) = \eta \tag{40}$$

From an application of Stokes' theorem to Maxwell's curl equations, $\nabla \times \mathbf{E} = -\mathbf{J}_m + i\omega\mu\mathbf{H}$ and $\nabla \times \mathbf{H} = \mathbf{J} - i\omega\epsilon\mathbf{E}$, it follows that the tangential components of **E** and **H** are respectively discontinuous by an amount equal to the magnetic surface-current density $\mathbf{K}_m$ and the electric surface-current density **K** in amperes per meter:

$$\mathbf{n} \times (\mathbf{E}_2 - \mathbf{E}_1) = -\mathbf{K}_m \qquad \mathbf{n} \times (\mathbf{H}_2 - \mathbf{H}_1) = \mathbf{K} \tag{41}$$

In these relations $\mathbf{K}_m$ and **K** are magnetic and electric "current sheets" carrying charge densities η_m and η respectively. Such current sheets are mathematical abstractions which can be simulated by limiting forms of electromagnetic objects. For example, if medium 1 is a perfect conductor and medium 2 a perfect dielectric, i.e., if $\sigma_1 = \infty$ and $\sigma_2 = 0$, then all the field vectors in medium 1 as well as η_m and $\mathbf{K}_m$ vanish identically and the boundary conditions reduce to

$$\mathbf{n} \cdot \mathbf{B}_2 = 0 \qquad \mathbf{n} \cdot \mathbf{D}_2 = \eta \qquad \mathbf{n} \times \mathbf{E}_2 = 0 \qquad \mathbf{n} \times \mathbf{H}_2 = \mathbf{K} \tag{42}$$

A surface having these boundary conditions is said to be an "electric wall." By duality a surface displaying the boundary conditions

$$\mathbf{n} \cdot \mathbf{D}_2 = 0 \qquad \mathbf{n} \cdot \mathbf{B}_2 = \eta_m \qquad \mathbf{n} \times \mathbf{E}_2 = -\mathbf{K}_m \qquad \mathbf{n} \times \mathbf{H}_2 = 0 \quad (43)$$

is said to be a "magnetic wall."

At sharp edges the field vectors may become infinite. However, the order of this singularity is restricted by the Bouwkamp-Meixner[1] edge condition. According to this condition, the energy density must be integrable over any finite domain even if this domain happens to include field singularities, i.e., the energy in any finite region of space must be finite. For example, when applied to a perfectly conducting sharp edge, this condition states that the singular components of the electric and magnetic vectors are of the order $\delta^{-1/2}$, where δ is the distance from the edge, whereas the parallel components are always finite.

1.4 The Field Potentials and Antipotentials

According to Helmholtz's partition theorem[2] any well-behaved vector field can be split into an irrotational part and a solenoidal part, or, equivalently, a vector field is determined by a knowledge of its curl and divergence. To partition an electromagnetic field generated by a current $\mathbf{J}$ and a charge ρ, we recall Maxwell's equations

$$\nabla \times \mathbf{H} = \mathbf{J} - i\omega\mathbf{D} \tag{44}$$

$$\nabla \times \mathbf{E} = i\omega\mathbf{B} \tag{45}$$

[1] C. Bouwkamp, *Physica,* **12:** 467 (1946); J. Meixner, *Ann. Phys.,* (6) **6:** 1 (1949).

[2] H. von Helmholtz, Über Integrale der hydrodynamischen Gleichungen, welche den Wirbelbewegungen entsprechen, *Crelles J.,* **55:** 25 (1858). This theorem was proved earlier in less complete form by G. B. Stokes in his paper On the Dynamical Theory of Diffraction, *Trans. Cambridge Phil. Soc.,* **9:** 1 (1849). For a mathematically rigorous proof, see O. Blumenthal, Uber die Zerlegung unendlicher Vektorfelder, *Math. Ann.,* **61:** 235 (1905).

$$\nabla \cdot \mathbf{D} = \rho \tag{46}$$

$$\nabla \cdot \mathbf{B} = 0 \tag{47}$$

and the constitutive relations for a simple medium

$$\mathbf{D} = \epsilon \mathbf{E} \tag{48}$$

$$\mathbf{B} = \mu \mathbf{H} \tag{49}$$

From the solenoidal nature of **B**, which is displayed by Eq. (47), it follows that **B** is derivable from a magnetic vector potential **A**:

$$\mathbf{B} = \nabla \times \mathbf{A} \tag{50}$$

This relation involves only the curl of **A** and leaves free the divergence of **A**. That is, $\nabla \cdot \mathbf{A}$ is not restricted and may be chosen arbitrarily to suit the needs of calculation. Inserting Eq. (50) into Eq. (45) we see that $\mathbf{E} - i\omega\mathbf{A}$ is irrotational and hence derivable from a scalar electric potential ϕ:

$$\mathbf{E} = -\nabla\phi + i\omega\mathbf{A} \tag{51}$$

This expression does not necessarily constitute a complete partition of the electric field because **A** itself may possess both irrotational and solenoidal parts. Only when **A** is purely solenoidal is the electric field completely partitioned into an irrotational part $\nabla\phi$ and a solenoidal part **A**. The magnetic field need not be partitioned intentionally because it is always purely solenoidal.

By virtue of their form, expressions (50) and (51) satisfy the two Maxwell equations (45) and (47). But in addition they must also satisfy the other two Maxwell equations, which, with the aid of the constitutive relations (48) and (49), become

$$\frac{1}{\mu} \nabla \times \mathbf{B} = \mathbf{J} - i\omega\epsilon\mathbf{E} \qquad \text{and} \qquad \nabla \cdot \mathbf{E} = \rho/\epsilon \tag{52}$$

When relations (50) and (51) are substituted into these equations, the following simultaneous differential equations are obtained,[1] relating ϕ and $\mathbf{A}$ to the source quantities $\mathbf{J}$ and ρ:

$$\nabla^2\phi - i\omega\nabla \cdot \mathbf{A} = -\rho/\epsilon \tag{53}$$

$$\nabla^2\mathbf{A} + k^2\mathbf{A} = -\mu\mathbf{J} + \nabla(\nabla \cdot \mathbf{A} - i\omega\epsilon\mu\phi) \tag{54}$$

where $k^2 = \omega^2\mu\epsilon$. Here $\nabla \cdot \mathbf{A}$ is not yet specified and may be chosen to suit our convenience. Clearly a prudent choice is one that uncouples the equations, i.e., reduces the system to an equation involving ϕ alone and an equation involving $\mathbf{A}$ alone. Accordingly, we choose $\nabla \cdot \mathbf{A} = i\omega\epsilon\mu\phi$ or $\nabla \cdot \mathbf{A} = 0$.

If we choose the Lorentz gauge

$$\nabla \cdot \mathbf{A} = i\omega\epsilon\mu\phi \tag{55}$$

then Eqs. (53) and (54) reduce to the Helmholtz equations

$$\nabla^2\phi + k^2\phi = -\rho/\epsilon \tag{56}$$

$$\nabla^2\mathbf{A} + k^2\mathbf{A} = -\mu\mathbf{J} \tag{57}$$

The Lorentz gauge is the conventional one, but in this gauge the electric field is not completely partitioned. If complete partition is desired, we must choose the Coulomb gauge[2]

$$\nabla \cdot \mathbf{A} = 0 \tag{58}$$

[1] Also the vector identity $\nabla \times \nabla \times \mathbf{A} = \nabla(\nabla \cdot \mathbf{A}) - \nabla^2\mathbf{A}$ is used. The quantity $\nabla^2\mathbf{A}$ is defined by the identity itself or by the formal operation $\nabla^2\mathbf{A} = \sum_i \nabla^2(\mathbf{e}_i A_i)$, where the A_i are the components of $\mathbf{A}$ and the $\mathbf{e}_i$ are the unit base vectors of the coordinate system. The Laplacian ∇^2 operates on not only the A_i but also the $\mathbf{e}_i$. In the special case of cartesian coordinates, the base vectors are constant; hence the Laplacian operates on only the A_i, that is, $\nabla^2\mathbf{A} = \sum_i \mathbf{e}_i\nabla^2 A_i$. See, for example, P. M. Morse and H. Feshbach, "Methods of Theoretical Physics," part I, pp. 51–52, McGraw-Hill Book Company, New York, 1953.

[2] See, for example, W. R. Smythe, "Static and Dynamic Electricity," 2d ed., p. 469, McGraw-Hill Book Company, New York, 1950.

which reduces Eqs. (53) and (54) to

$$\nabla^2\phi = -\frac{1}{\epsilon}\rho \tag{59}$$

$$\nabla^2\mathbf{A} + k^2\mathbf{A} = -\mu\mathbf{J} - i\omega\epsilon\mu\nabla\phi \tag{60}$$

We note that Eq. (59) is Poisson's equation and can be reduced no further. However, Eq. (60) may be simplified by partitioning $\mathbf{J}$ into an irrotational part $\mathbf{J}_i$ and a solenoidal part $\mathbf{J}_s$, and by noting that the irrotational part just cancels the term involving the gradient. To show this, $\mathbf{J}$ is split up as follows: $\mathbf{J} = \mathbf{J}_i + \mathbf{J}_s$, where by definition $\nabla \times \mathbf{J}_i = 0$ and $\nabla \cdot \mathbf{J}_s = 0$. Since $\mathbf{J}_i$ is irrotational, it is derivable from a scalar function ψ, viz., $\mathbf{J}_i = \nabla\psi$. The divergence of this relation, $\nabla \cdot \mathbf{J}_i = \nabla^2\psi$, when combined with the continuity equation $\nabla \cdot \mathbf{J} = \nabla \cdot (\mathbf{J}_i + \mathbf{J}_s) = \nabla \cdot \mathbf{J}_i = i\omega\rho$, leads to $\nabla^2\psi = i\omega\rho$. A comparison of this result with Eq. (59) shows that $\psi = -i\omega\epsilon\phi$ and hence $\mathbf{J}_i = \nabla\psi = -i\omega\epsilon\nabla\phi$. From this expression it therefore follows that $-\mu\mathbf{J}_i - i\omega\epsilon\mu\nabla\phi$ vanishes and consequently Eq. (60) reduces to

$$\nabla^2\mathbf{A} + k^2\mathbf{A} = -\mu\mathbf{J}_s \tag{61}$$

Thus we see that in this gauge, $\mathbf{A}$ is determined by the solenoidal part $\mathbf{J}_s$ of the current distribution and ϕ by its irrotational part $\mathbf{J}_i$. Since ϕ satisfies Poisson's equation, its spatial distribution resembles that of an electrostatic potential and therefore contributes predominantly to the near-zone electric field. It is like an electrostatic field only in its space dependence; its time dependence is harmonic.

In regions free of current ($\mathbf{J} = 0$) and charge ($\rho = 0$) we may supplement the gauge $\nabla \cdot \mathbf{A} = 0$ by taking $\phi \equiv 0$. Then Eq. (53) is trivially satisfied and Eq. (54) reduces to the homogeneous Helmholtz equation

$$\nabla^2\mathbf{A} + k^2\mathbf{A} = 0 \tag{62}$$

In this case the electromagnetic field is derived from the vector potential $\mathbf{A}$ alone.

Let us now partition the electromagnetic field generated by a magnetic current $\mathbf{J}_m$ and a magnetic charge ρ_m. We recall that Maxwell's equations for such a field are

$$\nabla \times \mathbf{H} = -i\omega \mathbf{D} \tag{63}$$

$$\nabla \times \mathbf{E} = -\mathbf{J}_m + i\omega \mathbf{B} \tag{64}$$

$$\nabla \cdot \mathbf{D} = 0 \tag{65}$$

$$\nabla \cdot \mathbf{B} = \rho_m \tag{66}$$

and, as before, the constitutive relations (48) and (49) are valid. From Eq. (65) it follows that $\mathbf{D}$ is solenoidal and hence derivable from an electric vector potential $\mathbf{A}_e$:

$$\mathbf{D} = -\nabla \times \mathbf{A}_e \tag{67}$$

In turn it follows from Eq. (63) that $\mathbf{H} - i\omega\mathbf{A}_e$ is irrotational and hence equal to $-\nabla\phi_m$, where ϕ_m is a magnetic scalar potential:

$$\mathbf{H} = -\nabla\phi_m + i\omega\mathbf{A}_e \tag{68}$$

Substituting expressions (67) and (68) into Eqs. (64) and (66), we get, with the aid of the constitutive relations, the following differential equations for $\mathbf{A}_e$ and ϕ_m:

$$\begin{aligned} \nabla^2\phi_m - i\omega\nabla \cdot \mathbf{A}_e &= -\frac{1}{\mu}\rho_m \\ \nabla^2\mathbf{A}_e + k^2\mathbf{A}_e &= -\epsilon\mathbf{J}_m + \nabla(\nabla \cdot \mathbf{A}_e - i\omega\mu\epsilon\phi_m) \end{aligned} \tag{69}$$

If we choose the conventional gauge

$$\nabla \cdot \mathbf{A}_e = i\omega\mu\epsilon\phi_m \tag{70}$$

then ϕ_m and $\mathbf{A}_e$ satisfy

$$\nabla^2\phi_m + k^2\phi_m = -\frac{1}{\mu}\rho_m \tag{71}$$

$$\nabla^2\mathbf{A}_e + k^2\mathbf{A}_e = -\epsilon\mathbf{J}_m \tag{72}$$

In this gauge ϕ_m and $\mathbf{A}_e$ are called "antipotentials." Clearly we may also choose the gauge $\nabla \cdot \mathbf{A}_e = 0$ which leads to

$$\nabla^2 \phi_m = -\frac{1}{\mu} \rho_m$$
$$\nabla^2 \mathbf{A}_e + k^2 \mathbf{A}_e = -\epsilon \mathbf{J}_{ms} \tag{73}$$

where $\mathbf{J}_{ms}$ is the solenoidal part of the magnetic current; this gauge leads also to $\phi_m = 0$ and

$$\nabla^2 \mathbf{A}_e + k^2 \mathbf{A}_e = 0 \tag{74}$$

for regions where $\mathbf{J}_m = 0$ and $\rho_m = 0$.

If the electromagnetic field is due to magnetic as well as electric currents and charges, then the field for the conventional gauge is given in terms of the potentials $\mathbf{A}$, ϕ and the antipotentials $\mathbf{A}_e$, ϕ_m by

$$\mathbf{E} = -\nabla \phi + i\omega \mathbf{A} - \frac{1}{\epsilon} \nabla \times \mathbf{A}_e \tag{75}$$

$$\mathbf{B} = \nabla \times \mathbf{A} - \mu \nabla \phi_m + i\omega\mu \mathbf{A}_e \tag{76}$$

1.5 Energy Relations

The instantaneous electric and magnetic energy densities for a lossless medium are defined respectively by

$$w_e = \int \mathbf{E}(t) \cdot \frac{\partial}{\partial t} \mathbf{D}(t) dt \qquad \text{and} \qquad w_m = \int \mathbf{H}(t) \cdot \frac{\partial}{\partial t} \mathbf{B}(t) dt \tag{77}$$

where $\mathbf{E}(t)$ stands for $\mathbf{E}(\mathbf{r},t)$, $\mathbf{D}(t)$ for $\mathbf{D}(\mathbf{r},t)$, etc. In the present instance these expressions reduce to

$$w_e = \tfrac{1}{2}\epsilon \mathbf{E}(t) \cdot \mathbf{E}(t) \qquad \text{and} \qquad w_m = \tfrac{1}{2}\mu \mathbf{H}(t) \cdot \mathbf{H}(t) \tag{78}$$

Both w_e and w_m are measured in joules per meter^3. To transform these quadratic quantities into the monochromatic domain we recall that

$$\mathbf{E}(t) = \text{Re}\,\{\mathbf{E}e^{-i\omega t}\} \qquad \text{and} \qquad \mathbf{H}(t) = \text{Re}\,\{\mathbf{H}e^{-i\omega t}\} \tag{79}$$

where $\mathbf{E}$ is shorthand for $\mathbf{E}_\omega(\mathbf{r})$ and $\mathbf{H}$ for $\mathbf{H}_\omega(\mathbf{r})$. Since $\mathbf{E}$ can always be written as $\mathbf{E} = \mathbf{E}_1 + i\mathbf{E}_2$, where $\mathbf{E}_1$ and $\mathbf{E}_2$ are respectively the real and imaginary parts of $\mathbf{E}$, the first of Eqs. (79) is equivalent to

$$\mathbf{E}(t) = \mathbf{E}_1 \cos \omega t + \mathbf{E}_2 \sin \omega t \tag{80}$$

Inserting this representation into the first of Eqs. (78) we obtain

$$w_e = \tfrac{1}{2}\epsilon\mathbf{E}_1 \cdot \mathbf{E}_1 \cos^2 \omega t + \tfrac{1}{2}\epsilon\mathbf{E}_2 \cdot \mathbf{E}_2 \sin^2 \omega t + \epsilon\mathbf{E}_1 \cdot \mathbf{E}_2 \sin \omega t \cos \omega t \tag{81}$$

which, when averaged over a period, yields the time-average electric energy density

$$\bar{w}_e = \tfrac{1}{4}\epsilon\mathbf{E}_1 \cdot \mathbf{E}_1 + \tfrac{1}{4}\epsilon\mathbf{E}_2 \cdot \mathbf{E}_2 \tag{82}$$

where the bar denotes the time average. Since

$$\mathbf{E}_1 \cdot \mathbf{E}_1 + \mathbf{E}_2 \cdot \mathbf{E}_2 = \mathbf{E} \cdot \mathbf{E}^*$$

where $\mathbf{E}^*$ is the conjugate complex of $\mathbf{E}$, we can express $\bar{w}_e$ in the equivalent form

$$w_e = \tfrac{1}{4}\epsilon\mathbf{E} \cdot \mathbf{E}^* \tag{83}$$

By a similar procedure it follows from the second of Eqs. (78) and the second of Eqs. (79) that the time-average magnetic energy is given by

$$w_m = \tfrac{1}{4}\mu\mathbf{H} \cdot \mathbf{H}^* \tag{84}$$

The instantaneous Poynting vector $\mathbf{S}(t)$ is defined by

$$\mathbf{S}(t) = \mathbf{E}(t) \times \mathbf{H}(t) \tag{85}$$

where $\mathbf{S}(t)$ stands for $\mathbf{S}(\mathbf{r},t)$ and is measured in watts per meter². With the aid of expressions (79), the time average of Eq. (85) leads to the following expression for the complex Poynting vector:

$$\mathbf{S} = \tfrac{1}{2}\mathbf{E} \times \mathbf{H}^* \tag{86}$$

If from the scalar product of $\mathbf{H}^*$ and $\nabla \times \mathbf{E} = i\omega\mu\mathbf{H}$ the scalar product of $\mathbf{E}$ and $\nabla \times \mathbf{H}^* = \mathbf{J}^* + i\omega\epsilon\mathbf{E}^*$($\epsilon$ is assumed to be real) is subtracted, and if use is made of the vector identity

$$\nabla \cdot (\mathbf{E} \times \mathbf{H}^*) = \mathbf{H}^* \cdot \nabla \times \mathbf{E} - \mathbf{E} \cdot \nabla \times \mathbf{H}^*$$

the following equation is obtained:

$$\nabla \cdot (\mathbf{E} \times \mathbf{H}^*) = -\mathbf{J}^* \cdot \mathbf{E} + i\omega(\mu\mathbf{H} \cdot \mathbf{H}^* - \epsilon\mathbf{E} \cdot \mathbf{E}^*) \tag{87}$$

which, with the aid of definitions (83), (84), and (86), yields the monochromatic form of Poynting's vector theorem[1]

$$\nabla \cdot \mathbf{S} = -\tfrac{1}{2}\mathbf{J}^* \cdot \mathbf{E} + 2i\omega(\bar{w}_m - \bar{w}_e) \tag{88}$$

The real part of this relation, i.e.,

$$\nabla \cdot (\mathrm{Re}\,\mathbf{S}) = \mathrm{Re}\,(-\tfrac{1}{2}\mathbf{J}^* \cdot \mathbf{E}) \tag{89}$$

expresses the conservation of time-average power, the term on the right representing a source (when positive) or a sink (when negative) and correspondingly the one on the left an outflow (when positive) or an inflow (when negative).

In Poynting's vector theorem (88) a term involving the difference $\bar{w}_m - \bar{w}_e$ appears. To obtain an energy relation (for the monochromatic state) which contains the sum $\bar{w}_m + \bar{w}_e$ instead of the difference $\bar{w}_m - \bar{w}_e$ we proceed as follows. From vector analysis we recall that the quantity

$$\nabla \cdot \left(\frac{\partial \mathbf{E}}{\partial \omega} \times \mathbf{H}^* + \mathbf{E}^* \times \frac{\partial \mathbf{H}}{\partial \omega}\right) \tag{90}$$

[1] F. Emde, *Elektrotech. Maschinenbau*, **27:** 112 (1909).

is identically equal to

$$\mathbf{H}^* \cdot \nabla \times \frac{\partial \mathbf{E}}{\partial \omega} - \frac{\partial \mathbf{E}}{\partial \omega} \cdot \nabla \times \mathbf{H}^* + \frac{\partial \mathbf{H}}{\partial \omega} \cdot \nabla \times \mathbf{E}^* - \mathbf{E}^* \cdot \nabla \times \frac{\partial \mathbf{H}}{\partial \omega} \tag{91}$$

From Maxwell's equations $\nabla \times \mathbf{E} = i\omega\mu\mathbf{H}$ and $\nabla \times \mathbf{H} = \mathbf{J} - i\omega\epsilon\mathbf{E}$ it follows that

$$\nabla \times \frac{\partial \mathbf{E}}{\partial \omega} = \frac{\partial}{\partial \omega}(\nabla \times \mathbf{E}) = \frac{\partial}{\partial \omega}(i\omega\mu\mathbf{H}) = i\mu\mathbf{H} + i\omega\mu\frac{\partial \mathbf{H}}{\partial \omega} + i\omega\mathbf{H}\frac{\partial \mu}{\partial \omega}$$

$$\nabla \times \frac{\partial \mathbf{H}}{\partial \omega} = \frac{\partial}{\partial \omega}(\nabla \times \mathbf{H}) = \frac{\partial}{\partial \omega}(\mathbf{J} - i\omega\epsilon\mathbf{E}) = \frac{\partial \mathbf{J}}{\partial \omega} - i\epsilon\mathbf{E} - i\omega\epsilon\frac{\partial \mathbf{E}}{\partial \omega} - i\omega\mathbf{E}\frac{\partial \epsilon}{\partial \omega}$$

$$\nabla \times \mathbf{E}^* = -i\omega\mu\mathbf{H}^* \qquad \text{and} \qquad \nabla \times \mathbf{H}^* = \mathbf{J}^* + i\omega\epsilon\mathbf{E}^*$$

Substituting these relations into expression (91) we obtain the desired energy relation

$$\nabla \cdot \left(\frac{\partial \mathbf{E}}{\partial \omega} \times \mathbf{H}^* + \mathbf{E}^* \times \frac{\partial \mathbf{H}}{\partial \omega}\right) = i\left[\frac{\partial(\omega\mu)}{\partial \omega}\mathbf{H} \cdot \mathbf{H}^* + \frac{\partial(\omega\epsilon)}{\partial \omega}\mathbf{E} \cdot \mathbf{E}^*\right] - \frac{\partial \mathbf{E}}{\partial \omega} \cdot \mathbf{J}^* - \mathbf{E}^* \cdot \frac{\partial \mathbf{J}}{\partial \omega} \tag{92}$$

which we call the "energy theorem." Here we interpret as the time-average electric and magnetic energy densities the quantities

$$\bar{w}_e = \frac{1}{4}\frac{\partial(\omega\epsilon)}{\partial \omega}\mathbf{E} \cdot \mathbf{E}^* \qquad \bar{w}_m = \frac{1}{4}\frac{\partial(\omega\mu)}{\partial \omega}\mathbf{H} \cdot \mathbf{H}^* \tag{93}$$

which reduce respectively to expressions (83) and (84) when the medium is nondispersive, i.e., when $\partial\epsilon/\partial\omega = 0$ and $\partial\mu/\partial\omega = 0$.

Radiation from monochromatic sources in unbounded regions

2

The problem of determining the electromagnetic field radiated by a given monochromatic source in a simple, unbounded medium is usually handled by first finding the potentials of the source and then calculating the field from a knowledge of these potentials. However, this is not the only method of determining the field. There is an alternative method, that of the dyadic Green's function, which yields the field directly in terms of the source current. In this chapter these two methods are discussed.

2.1 The Helmholtz Integrals

We wish to find the vector potential **A** and the scalar potential ϕ of a monochromatic current **J**, which is confined to a region of finite spatial extent and completely surrounded by a simple, lossless, unbounded medium. For this purpose it is convenient to choose the Lorentz gauge

$$\nabla \cdot \mathbf{A} = i\omega\epsilon\mu\phi \tag{1}$$

In this gauge, ϕ and **A** must satisfy the Helmholtz equations (see Sec. 1.4)

$$\nabla^2\phi(\mathbf{r}) + k^2\phi(\mathbf{r}) = -\frac{1}{\epsilon}\rho(\mathbf{r}) \tag{2}$$

$$\nabla^2\mathbf{A}(\mathbf{r}) + k^2\mathbf{A}(\mathbf{r}) = -\mu\mathbf{J}(\mathbf{r}) \tag{3}$$

Since the medium is unbounded, ϕ and $\mathbf{A}$ must also satisfy the radiation condition. In physical terms this means that ϕ and $\mathbf{A}$ in the far zone must have the form of outwardly traveling spherical (but not necessarily isotropic) waves, the sphericity of the waves being a consequence of the confinement of the sources ρ and $\mathbf{J}$ to a finite part of space.

Let us first consider the problem of finding ϕ. We recall from the theory of the scalar Helmholtz equation that ϕ is uniquely determined by Eq. (2) and by the radiation condition[1]

$$\lim_{r\to\infty} r\left(\frac{\partial\phi}{\partial r} - ik\phi\right) = 0 \tag{4}$$

where $r = (\sqrt{\mathbf{r}\cdot\mathbf{r}})$ is the radial coordinate of a spherical coordinate system r, θ, ψ. To deduce from this radiation condition the explicit behavior of ϕ on the sphere at infinity, we note that the scalar Helmholtz equation is separable in spherical coordinates and then write ϕ in the separated form $\phi(\mathbf{r}) = f(\theta,\psi)\, u(r)$, where f is a function of the angular coordinates and u is a function of r only. Clearly the radiation condition (4) is satisfied by $u(r) = (1/r) \exp(ikr)$ and accordingly at great distances from the source the behavior of ϕ must be in accord with

$$\lim_{r\to\infty} \phi(\mathbf{r}) = f(\theta,\psi)\,\frac{e^{ikr}}{r} \tag{5}$$

That is, the solution of Eq. (2) that we are seeking is the one that has the far-zone behavior (5).

Since the scalar Helmholtz equation (2) is linear, we may write ϕ in the form[2]

$$\phi(\mathbf{r}) = \frac{1}{\epsilon}\int \rho(\mathbf{r}')G(\mathbf{r},\mathbf{r}')dV' \tag{6}$$

[1] This is Sommerfeld's "Ausstrahlungbedingung"; see A. Sommerfeld, Die Greensche Funktion der Schwingungsgleichung, *Jahresbericht d. D. Math. Ver.*, **21:** 309 (1912).

[2] From the point of view of the theory of differential equations, the solution of Eq. (2) consists of not only the particular integral (6) but also a complementary solution. In the present instance, however, the radiation condition requires that the complementary solution vanish identically.

where $G(\mathbf{r},\mathbf{r}')$ is a function of the coordinates of the observation point $\mathbf{r}$ and of the source point $\mathbf{r}'$, and where the integration with respect to the primed coordinates extends throughout the volume V occupied by ρ. The unknown function G is determined by making expression (6) satisfy Eq. (2) and condition (5). Substituting expression (6) into Eq. (2) we get

$$\int \rho(\mathbf{r}')(\nabla^2 + k^2)G(\mathbf{r},\mathbf{r}')dV' = -\rho(\mathbf{r}) \tag{7}$$

where the Laplacian operator operates with respect to the unprimed coordinates only. Then with the aid of the Dirac δ function[1] which permits ρ to be represented as the volume integral

$$\rho(\mathbf{r}) = \int \rho(\mathbf{r}')\delta(\mathbf{r} - \mathbf{r}')dV' \qquad (\mathbf{r} \text{ in } V) \tag{8}$$

we see that Eq. (7) can be written as

$$\int \rho(\mathbf{r}')[(\nabla^2 + k^2)G(\mathbf{r},\mathbf{r}') + \delta(\mathbf{r} - \mathbf{r}')]dV' = 0 \tag{9}$$

From this it follows that G must satisfy the scalar Helmholtz equation

$$\nabla^2 G(\mathbf{r},\mathbf{r}') + k^2 G(\mathbf{r},\mathbf{r}') = -\delta(\mathbf{r} - \mathbf{r}') \tag{10}$$

Since G satisfies Eq. (2) with its source term replaced by a δ function, G is said to be a Green's function[2] of Eq. (2).

The appropriate solution of Eq. (10) for $\mathbf{r} \neq \mathbf{r}'$ is

$$G(\mathbf{r},\mathbf{r}') = \alpha \frac{e^{ik|\mathbf{r}-\mathbf{r}'|}}{|\mathbf{r} - \mathbf{r}'|} \tag{11}$$

[1] The δ function has the following definitive properties: $\delta(\mathbf{r} - \mathbf{r}') = 0$ for $\mathbf{r} \neq \mathbf{r}'$ and $= \infty$ for $\mathbf{r} = \mathbf{r}'$; $\int_v f(\mathbf{r})\delta(\mathbf{r} - \mathbf{r}')dV = f(\mathbf{r}')$ for $\mathbf{r}'$ in V and $= 0$ for $\mathbf{r}'$ outside of V where f is any well-behaved function. See P. A. M. Dirac, "The Principles of Quantum Mechanics," pp. 58–61, Oxford University Press, London, 1947. See also L. Schwartz, Théorie des distributions, *Actualités scientifiques et industrielles*, **1091** and **1122,** Hermann et Cie, Paris, 1950–51.

[2] See, for example, R. Courant and D. Hilbert, "Methods of Mathematical Physics," vol. 1, pp. 351–388, Interscience Publishers, Inc., New York, 1953.

where α is a constant. It becomes clear that this solution is compatible with the requirement that the form (6) satisfy condition (5) when we recall the geometric relation

$$|\mathbf{r} - \mathbf{r}'| = \sqrt{r^2 + r'^2 - 2\mathbf{r} \cdot \mathbf{r}'} = r\sqrt{1 + (r'/r)^2 - 2\mathbf{r}' \cdot \mathbf{r}/r^2} \tag{12}$$

where $r^2 = \mathbf{r} \cdot \mathbf{r}$ and $r'^2 = \mathbf{r}' \cdot \mathbf{r}'$, and from this relation find the limiting form

$$\lim_{r\to\infty} G(\mathbf{r},\mathbf{r}') = \lim_{r\to\infty} \alpha \frac{e^{ik|\mathbf{r}-\mathbf{r}'|}}{|\mathbf{r} - \mathbf{r}'|} \to \alpha \frac{e^{ikr}}{r} \exp(-ik\mathbf{r}' \cdot \mathbf{r}/r) \tag{13}$$

To determine the constant α, expression (11) is substituted into Eq. (10) and the resulting equation is integrated throughout a small spherical volume centered on the point $\mathbf{r} = \mathbf{r}'$. It turns out that α must be equal to $1/4\pi$, and hence the Green's function is

$$G(\mathbf{r},\mathbf{r}') = \frac{e^{ik|\mathbf{r}-\mathbf{r}'|}}{4\pi|\mathbf{r} - \mathbf{r}'|} \tag{14}$$

Therefore, since the form (6) satisfies Eq. (2) and condition (5) when G is given by expression (14), the desired solution of Eq. (2) can be written as the Helmholtz integral

$$\phi(\mathbf{r}) = \frac{1}{\epsilon} \int \rho(\mathbf{r}') \frac{e^{ik|\mathbf{r}-\mathbf{r}'|}}{4\pi|\mathbf{r} - \mathbf{r}'|} dV' \tag{15}$$

Now the related problem of finding $\mathbf{A}$ can be easily handled. Clearly, the appropriate solution of Eq. (3) must be the Helmholtz integral

$$\mathbf{A}(\mathbf{r}) = \mu \int \mathbf{J}(\mathbf{r}') \frac{e^{ik|\mathbf{r}-\mathbf{r}'|}}{4\pi|\mathbf{r} - \mathbf{r}'|} dV' \tag{16}$$

because it has the proper behavior on the sphere at infinity and it satisfies Eq. (3). To show that it satisfies Eq. (3), one only has to operate on Eq. (16) with the operator $(\nabla + k^2)$ and note that $\mathbf{J}(\mathbf{r}')$ depends on the primed coordinates alone and that the Green's function (14) obeys Eq. (10).

When in addition to the electric current $\mathbf{J}$ there is a monochromatic magnetic current distribution $\mathbf{J}_m$ of finite spatial extent, the antipotentials ϕ_m and $\mathbf{A}_e$ should be invoked. The magnetic scalar potential ϕ_m and the electric vector potential $\mathbf{A}_e$ satisfy the Helmholtz equations

(see Sec. 1.4)

$$\nabla^2\phi_m(\mathbf{r}) + k^2\phi_m(\mathbf{r}) = -\frac{1}{\mu}\rho_m(\mathbf{r}) \tag{17}$$

$$\nabla^2\mathbf{A}_e(\mathbf{r}) + k^2\mathbf{A}_e(\mathbf{r}) = -\epsilon\mathbf{J}_m(\mathbf{r}) \tag{18}$$

where $\nabla \cdot \mathbf{A}_e = i\omega\mu\epsilon\phi_m$ and $\nabla \cdot \mathbf{J}_m = i\omega\rho_m$. A procedure similar to the one we used in obtaining the Helmholtz integrals for ϕ and $\mathbf{A}$ leads to the following Helmholtz integrals for ϕ_m and $\mathbf{A}_e$:

$$\phi_m(\mathbf{r}) = \frac{1}{\mu}\int \rho_m(\mathbf{r}')\,\frac{e^{ik|\mathbf{r}-\mathbf{r}'|}}{4\pi|\mathbf{r}-\mathbf{r}'|}\,dV' \tag{19}$$

$$\mathbf{A}_e(\mathbf{r}) = \epsilon\int \mathbf{J}_m(\mathbf{r}')\,\frac{e^{ik|\mathbf{r}-\mathbf{r}'|}}{4\pi|\mathbf{r}-\mathbf{r}'|}\,dV' \tag{20}$$

From a knowledge of ϕ, $\mathbf{A}$, ϕ_m, $\mathbf{A}_e$ the radiated electric and magnetic fields can be derived by use of the relations (see Sec. 1.4)

$$\mathbf{E} = -\nabla\phi + i\omega\mathbf{A} - \frac{1}{\epsilon}\nabla \times \mathbf{A}_e \tag{21}$$

$$\mathbf{H} = \frac{1}{\mu}\nabla \times \mathbf{A} - \nabla\phi_m + i\omega\mathbf{A}_e \tag{22}$$

It is sometimes desirable to eliminate ϕ and ϕ_m from these relations and thereby express $\mathbf{E}$ and $\mathbf{H}$ in terms of $\mathbf{A}$ and $\mathbf{A}_e$ only. This can be done with the aid of

$$-\nabla\phi = \frac{i}{\omega\epsilon\mu}\nabla(\nabla \cdot \mathbf{A}) \qquad \text{and} \qquad -\nabla\phi_m = \frac{i}{\omega\epsilon\mu}\nabla(\nabla \cdot \mathbf{A}_e) \tag{23}$$

which follow from the gradients of the Lorentz conditions $\nabla \cdot \mathbf{A} = i\omega\epsilon\mu\phi$ and $\nabla \cdot \mathbf{A}_e = i\omega\mu\phi_m$. Thus relations (21) and (22) may be written as follows:

$$\mathbf{E} = i\omega\left[\mathbf{A} + \frac{1}{k^2}\nabla(\nabla \cdot \mathbf{A})\right] - \frac{1}{\epsilon}\nabla \times \mathbf{A}_e \tag{24}$$

$$\mathbf{H} = \frac{1}{\mu}\nabla \times \mathbf{A} + i\omega\left[\mathbf{A}_e + \frac{1}{k^2}\nabla(\nabla \cdot \mathbf{A}_e)\right] \tag{25}$$

To enable us to cast $\mathbf{A} + \frac{1}{k^2}\nabla(\nabla \cdot \mathbf{A})$ and $\mathbf{A}_e + \frac{1}{k^2}\nabla(\nabla \cdot \mathbf{A}_e)$ into the form of an operator operating on $\mathbf{A}$ and $\mathbf{A}_e$, we introduce the unit dyadic $\mathbf{u}$ and the double-gradient dyadic $\nabla\nabla$ which in a cartesian system of coordinates are expressed by

$$\mathbf{u} = \sum_{m=1}^{m=3} \sum_{n=1}^{n=3} \mathbf{e}_m \mathbf{e}_n \delta_{mn} \tag{26}$$

$$\nabla\nabla = \sum_{m=1}^{m=3} \sum_{n=1}^{n=3} \mathbf{e}_m \mathbf{e}_n \frac{\partial}{\partial x_m} \frac{\partial}{\partial x_n} \tag{27}$$

where x_i ($i = 1, 2, 3$) are the cartesian coordinates, $\mathbf{e}_i$ ($i = 1, 2, 3$) are the unit base vectors, and the symbol δ_{mn} is the Kronecker delta, which is 1 for $m = n$ and 0 for $m \neq n$. The properties of $\mathbf{u}$ and $\nabla\nabla$ that we will need are $\mathbf{u} \cdot \mathbf{C} = \mathbf{C}$ and $(\nabla\nabla) \cdot \mathbf{C} = \nabla(\nabla \cdot \mathbf{C})$, where $\mathbf{C}$ is any vector function. These properties can be demonstrated by writing $\mathbf{C}$ in component form and then carrying out the calculation. Thus

$$\begin{aligned}\mathbf{u} \cdot \mathbf{C} &= \Big(\sum_m \sum_n \mathbf{e}_m \mathbf{e}_n \delta_{mn}\Big) \cdot \Big(\sum_p \mathbf{e}_p C_p\Big) \\ &= \sum_m \sum_n \sum_p \mathbf{e}_m \mathbf{e}_n \cdot \mathbf{e}_p C_p \delta_{mn} = \sum_m \sum_n \sum_p \mathbf{e}_m \delta_{np} C_p \delta_{mn} = \sum_p \mathbf{e}_p C_p = \mathbf{C}\end{aligned} \tag{28}$$

where $\mathbf{e}_n \cdot \mathbf{e}_p = \delta_{np}$, and

$$\begin{aligned}(\nabla\nabla) \cdot \mathbf{C} &= \left(\sum_m \sum_n \mathbf{e}_m \mathbf{e}_n \frac{\partial}{\partial x_m} \frac{\partial}{\partial x_n}\right) \cdot \left(\sum_p \mathbf{e}_p C_p\right) \\ &= \sum_m \sum_n \sum_p \mathbf{e}_m \mathbf{e}_n \cdot \mathbf{e}_p \frac{\partial}{\partial x_m} \frac{\partial}{\partial x_n} C_p = \sum_m \sum_n \sum_p \mathbf{e}_m \delta_{np} \frac{\partial}{\partial x_m} \frac{\partial}{\partial x_n} C_p \\ &= \sum_m \mathbf{e}_m \frac{\partial}{\partial x_m} \left(\sum_p \frac{\partial}{\partial x_p} C_p\right) = \nabla(\nabla \cdot \mathbf{C})\end{aligned} \tag{29}$$

With the aid of these results, relations (24) and (25) become

$$\mathbf{E} = i\omega\left(\mathbf{u} + \frac{1}{k^2}\nabla\nabla\right) \cdot \mathbf{A} - \frac{1}{\epsilon}\nabla \times \mathbf{A}_e \tag{30}$$

$$\mathbf{H} = \frac{1}{\mu} \nabla \times \mathbf{A} + i\omega \left(\mathbf{u} + \frac{1}{k^2} \nabla\nabla \right) \cdot \mathbf{A}_e \tag{31}$$

Using the Helmholtz integrals (16) and (20) and taking the curl operator and the operator $\mathbf{u} + \frac{1}{k^2} \nabla\nabla$ under the integral sign, we get

$$\mathbf{E} = i\omega\mu \int \left(\mathbf{u} + \frac{1}{k^2} \nabla\nabla \right) \cdot \left[\mathbf{J}(\mathbf{r}') \frac{e^{ik|\mathbf{r}-\mathbf{r}'|}}{4\pi|\mathbf{r} - \mathbf{r}'|} \right] dV' - \int \nabla \times \left[\mathbf{J}_m(\mathbf{r}') \frac{e^{ik|\mathbf{r}-\mathbf{r}'|}}{4\pi|\mathbf{r} - \mathbf{r}'|} \right] dV' \tag{32}$$

$$\mathbf{H} = \int \nabla \times \left[\mathbf{J}(\mathbf{r}') \frac{e^{ik|\mathbf{r}-\mathbf{r}'|}}{4\pi|\mathbf{r} - \mathbf{r}'|} \right] dV' + i\omega\epsilon \int \left(\mathbf{u} + \frac{1}{k^2} \nabla\nabla \right) \cdot \left[\mathbf{J}_m(\mathbf{r}') \frac{e^{ik|\mathbf{r}-\mathbf{r}'|}}{4\pi|\mathbf{r} - \mathbf{r}'|} \right] dV' \tag{33}$$

To reduce these expressions we invoke the following considerations. If $\mathbf{a}$ is a vector function of the primed coordinates only and w is a scalar function of the primed and unprimed coordinates, then

$$(\nabla\nabla) \cdot (\mathbf{a}w) = \left(\sum_m \sum_n \mathbf{e}_m \mathbf{e}_n \frac{\partial^2}{\partial x_m \partial x_n} \right) \cdot \mathbf{a}w = \left(\sum_m \sum_n \mathbf{e}_m \mathbf{e}_n \frac{\partial^2 w}{\partial x_m \partial x_n} \right) \cdot \mathbf{a} = (\nabla\nabla w) \cdot \mathbf{a} \tag{34}$$

and

$$\nabla \times (\mathbf{a}w) = \left(\sum_m \mathbf{e}_m \frac{\partial}{\partial x_m} \right) \times \mathbf{a}w = \left(\sum_m \mathbf{e}_m \frac{\partial w}{\partial x_m} \right) \times \mathbf{a} = \nabla w \times \mathbf{a} \tag{35}$$

In view of identities (34) and (35), expressions (32) and (33) reduce to the following:

$$\mathbf{E} = i\omega\mu \int \left[\left(\mathbf{u} + \frac{1}{k^2} \nabla\nabla \right) \frac{e^{ik|\mathbf{r}-\mathbf{r}'|}}{4\pi|\mathbf{r} - \mathbf{r}'|} \right] \cdot \mathbf{J}(\mathbf{r}') dV' - \int \left(\nabla \frac{e^{ik|\mathbf{r}-\mathbf{r}'|}}{4\pi|\mathbf{r} - \mathbf{r}'|} \right) \times \mathbf{J}_m(\mathbf{r}') dV' \tag{36}$$

$$\mathbf{H} = \int \left(\nabla \frac{e^{ik|\mathbf{r}-\mathbf{r}'|}}{4\pi|\mathbf{r}-\mathbf{r}'|}\right) \times \mathbf{J}(\mathbf{r}')dV' + i\omega\epsilon \int \left[\left(\mathbf{u} + \frac{1}{k^2}\nabla\nabla\right)\frac{e^{ik|\mathbf{r}-\mathbf{r}'|}}{4\pi|\mathbf{r}-\mathbf{r}'|}\right] \cdot \mathbf{J}_m(\mathbf{r}')dV' \tag{37}$$

Since the quantity

$$G(\mathbf{r},\mathbf{r}') \equiv \frac{e^{ik|\mathbf{r}-\mathbf{r}'|}}{4\pi|\mathbf{r}-\mathbf{r}'|} \tag{38}$$

is known as the free-space scalar Green's function, it is appropriate to refer to the quantity

$$\mathbf{\Gamma}(\mathbf{r},\mathbf{r}') \equiv \left(\mathbf{u} + \frac{1}{k^2}\nabla\nabla\right)\frac{e^{ik|\mathbf{r}-\mathbf{r}'|}}{4\pi|\mathbf{r}-\mathbf{r}'|} \equiv \left(\mathbf{u} + \frac{1}{k^2}\nabla\nabla\right)G(\mathbf{r},\mathbf{r}') \tag{39}$$

as the free-space dyadic Green's function. Using (38) and (39) we can write expressions (36) and (37) as follows:

$$\mathbf{E}(\mathbf{r}) = i\omega\mu\int\mathbf{\Gamma}(\mathbf{r},\mathbf{r}')\cdot\mathbf{J}(\mathbf{r}')dV' - \int\nabla G(\mathbf{r},\mathbf{r}')\times\mathbf{J}_m(\mathbf{r}')dV' \tag{40}$$

$$\mathbf{H}(\mathbf{r}) = \int\nabla G(\mathbf{r},\mathbf{r}')\times\mathbf{J}(\mathbf{r}')dV' + i\omega\epsilon\int\mathbf{\Gamma}(\mathbf{r},\mathbf{r}')\cdot\mathbf{J}_m(\mathbf{r}')dV' \tag{41}$$

These relations formally express the radiated fields **E**, **H** in terms of the source currents **J** and $\mathbf{J}_m$.[1]

2.2 Free-space Dyadic Green's Function

In the previous section we derived the free-space dyadic Green's function using the potentials and antipotentials as an intermediary. In this section we shall derive it directly from Maxwell's equations.

We denote the fields of the electric current by $\mathbf{E}'$, $\mathbf{H}'$ and those of the magnetic current by $\mathbf{E}''$, $\mathbf{H}''$. The resultant fields **E**, **H** are obtained

[1] If the point of observation **r** lies outside the region occupied by the source (which is the case of interest here), then $|\mathbf{r}-\mathbf{r}'| \neq 0$ everywhere and the integrals are proper. On the other hand, if **r** lies within the region of the source, then $|\mathbf{r}-\mathbf{r}'| = 0$ at one point in the region and there the integrals diverge. This improper behavior arises from interchanging the order of integration and differentiation. See, for example, J. Van Bladel, Some Remarks on Green's Dyadic for Infinite Space, *IRE Trans. Antennas Propagation*, **AP-9** (6): 563–566 (1961).

by superposition, i.e., $\mathbf{E} = \mathbf{E}' + \mathbf{E}''$ and $\mathbf{H} = \mathbf{H}' + \mathbf{H}''$. Let us consider first the fields $\mathbf{E}'$, $\mathbf{H}'$ which satisfy Maxwell's equations

$$\nabla \times \mathbf{H}' = \mathbf{J} - i\omega\epsilon\mathbf{E}' \qquad \text{and} \qquad \nabla \times \mathbf{E}' = i\omega\mu\mathbf{H}' \tag{42}$$

From these equations it follows that $\mathbf{E}'$ satisfies the vector Helmholtz equation with $\mathbf{J}$ as its source term:

$$\nabla \times \nabla \times \mathbf{E}' - k^2\mathbf{E}' = i\omega\mu\mathbf{J} \tag{43}$$

In this equation, $\mathbf{E}'$ is linearly related to $\mathbf{J}$; on the strength of this linearity we may write

$$\mathbf{E}'(\mathbf{r}) = i\omega\mu\int\mathbf{\Gamma}(\mathbf{r},\mathbf{r}') \cdot \mathbf{J}(\mathbf{r}')dV' \tag{44}$$

where $\mathbf{\Gamma}$ is an unknown dyadic function of $\mathbf{r}$ and $\mathbf{r}'$. To deduce the differential equation that $\mathbf{\Gamma}$ must satisfy we substitute this expression into the vector Helmholtz equation. Thus we obtain

$$\nabla \times \nabla \times \int\mathbf{\Gamma}(\mathbf{r},\mathbf{r}') \cdot \mathbf{J}(\mathbf{r}')dV' - k^2\int\mathbf{\Gamma}(\mathbf{r},\mathbf{r}') \cdot \mathbf{J}(\mathbf{r}')dV' = \int\mathbf{u} \cdot \mathbf{J}(\mathbf{r}')\delta(\mathbf{r} - \mathbf{r}')dV' \tag{45}$$

Noting that the double curl operator may be taken under the integral sign and observing that $\nabla \times \nabla \times (\mathbf{\Gamma} \cdot \mathbf{J}) = (\nabla \times \nabla \times \mathbf{\Gamma}) \cdot \mathbf{J}$, we get the following equation:

$$\int[\nabla \times \nabla \times \mathbf{\Gamma}(\mathbf{r},\mathbf{r}') - k^2\mathbf{\Gamma}(\mathbf{r},\mathbf{r}') - \mathbf{u}\delta(\mathbf{r} - \mathbf{r}')] \cdot \mathbf{J}(\mathbf{r}')dV' = 0 \tag{46}$$

Since this equation holds for any current distribution $\mathbf{J}(\mathbf{r}')$, it follows that $\mathbf{\Gamma}(\mathbf{r},\mathbf{r}')$ must satisfy

$$(\text{curl curl} - k^2)\mathbf{\Gamma}(\mathbf{r},\mathbf{r}') = \mathbf{u}\delta(\mathbf{r} - \mathbf{r}') \tag{47}$$

Now we construct a dyadic function $\mathbf{\Gamma}$ such that Eq. (47) will be satisfied and expression (44) will have the proper behavior on the sphere at infinity. One way of doing this is to use the identity curl curl = grad div $- \nabla^2$ and write Eq. (47) in the form

$$(\nabla^2 + k^2)\mathbf{\Gamma}(\mathbf{r},\mathbf{r}') = -\mathbf{u}\delta(\mathbf{r} - \mathbf{r}') + \nabla\nabla \cdot \mathbf{\Gamma}(\mathbf{r},\mathbf{r}') \tag{48}$$

From Eq. (47) it follows that $\nabla \cdot \mathbf{\Gamma}(\mathbf{r},\mathbf{r}') = -\frac{1}{k^2}\nabla\delta(\mathbf{r} - \mathbf{r}')$. With the aid of this relation, Eq. (48) becomes

$$(\nabla^2 + k^2)\mathbf{\Gamma}(\mathbf{r},\mathbf{r}') = -\left(\mathbf{u} + \frac{1}{k^2}\nabla\nabla\right)\delta(\mathbf{r} - \mathbf{r}') \tag{49}$$

Clearly this equation is satisfied by

$$\mathbf{\Gamma}(\mathbf{r},\mathbf{r}') = \left(\mathbf{u} + \frac{1}{k^2}\nabla\nabla\right)G(\mathbf{r},\mathbf{r}') \tag{50}$$

where $G(\mathbf{r},\mathbf{r}')$ in turn satisfies

$$(\nabla^2 + k^2)G(\mathbf{r},\mathbf{r}') = -\delta(\mathbf{r} - \mathbf{r}') \tag{51}$$

To meet the radiation condition, the solution of Eq. (51) must be

$$G(\mathbf{r},\mathbf{r}') = \frac{e^{ik|\mathbf{r}-\mathbf{r}'|}}{4\pi|\mathbf{r} - \mathbf{r}'|} \tag{52}$$

Thus the desired dyadic Green's function is

$$\mathbf{\Gamma}(\mathbf{r},\mathbf{r}') = \left(\mathbf{u} + \frac{1}{k^2}\nabla\nabla\right)\frac{e^{ik|\mathbf{r}-\mathbf{r}'|}}{4\pi|\mathbf{r} - \mathbf{r}'|} \tag{53}$$

The fields $\mathbf{E}''$, $\mathbf{H}''$ satisfy Maxwell's equations

$$\nabla \times \mathbf{H}'' = -i\omega\epsilon\mathbf{E}'' \qquad \text{and} \qquad \nabla \times \mathbf{E}'' = -\mathbf{J}_m + i\omega\mu\mathbf{H}'' \tag{54}$$

from which it follows that $\mathbf{H}''$ satisfies the vector Helmholtz equation with $\mathbf{J}_m$ as its source term:

$$\nabla \times \nabla \times \mathbf{H}'' - k^2\mathbf{H}'' = i\omega\epsilon\mathbf{J}_m \tag{55}$$

As before, if we write

$$\mathbf{H}'' = i\omega\epsilon\int\mathbf{\Gamma}(\mathbf{r},\mathbf{r}') \cdot \mathbf{J}_m(\mathbf{r}')dV' \tag{56}$$

then Eq. (55) and the radiation condition will be satisfied when $\mathbf{\Gamma}$ is given by expression (53). That is, the dyadic functions in the integrands of Eqs. (44) and (56) are identical.

Since $\mathbf{H}' = \frac{1}{i\omega\mu} \nabla \times \mathbf{E}'$ and $\mathbf{E}'' = -\frac{1}{i\omega\epsilon} \nabla \times \mathbf{H}''$, it follows from expressions (44) and (56) that

$$\mathbf{H}' = \nabla \times \int \mathbf{\Gamma}(\mathbf{r},\mathbf{r}') \cdot \mathbf{J}(\mathbf{r}')dV' = \int [\nabla \times \mathbf{\Gamma}(\mathbf{r},\mathbf{r}')] \cdot \mathbf{J}(\mathbf{r}')dV' \tag{57}$$

$$\mathbf{E}'' = -\nabla \times \int \mathbf{\Gamma}(\mathbf{r},\mathbf{r}') \cdot \mathbf{J}_m(\mathbf{r}')dV' = -\int [\nabla \times \mathbf{\Gamma}(\mathbf{r},\mathbf{r}')] \cdot \mathbf{J}_m(\mathbf{r}')dV' \tag{58}$$

But $\mathbf{\Gamma}(\mathbf{r},\mathbf{r}') = \left(\mathbf{u} + \frac{1}{k^2} \nabla\nabla\right) G(\mathbf{r},\mathbf{r}')$ and consequently

$$[\nabla \times \mathbf{\Gamma}(\mathbf{r},\mathbf{r}')] \cdot \mathbf{J}(\mathbf{r}') = \nabla G(\mathbf{r},\mathbf{r}') \times \mathbf{J}(\mathbf{r}')$$

In view of this, Eqs. (57) and (58) become

$$\mathbf{H}' = \int \nabla G(\mathbf{r},\mathbf{r}') \times \mathbf{J}(\mathbf{r}')dV' \tag{59}$$

$$\mathbf{E}'' = -\int \nabla G(\mathbf{r},\mathbf{r}') \times \mathbf{J}_m(\mathbf{r}')dV' \tag{60}$$

Combining expressions (44) and (56) with (60) and (59) respectively, we get

$$\mathbf{E} = \mathbf{E}' + \mathbf{E}'' = i\omega\mu \int \mathbf{\Gamma}(\mathbf{r},\mathbf{r}') \cdot \mathbf{J}(\mathbf{r}')dV' - \int \nabla G(\mathbf{r},\mathbf{r}') \times \mathbf{J}_m(\mathbf{r}')dV' \tag{61}$$

$$\mathbf{H} = \mathbf{H}' + \mathbf{H}'' = i\omega\epsilon \int \mathbf{\Gamma}(\mathbf{r},\mathbf{r}') \cdot \mathbf{J}_m(\mathbf{r}')dV' + \int \nabla G(\mathbf{r},\mathbf{r}') \times \mathbf{J}(\mathbf{r}')dV' \tag{62}$$

These expressions are identical to expressions (40) and (41).

2.3 Radiated Power

For the computation of the power radiated by a monochromatic electric current, the complex Poynting vector theorem (see Sec. 1.5)

$$\nabla \cdot \mathbf{S} = -\tfrac{1}{2}\mathbf{J}^* \cdot \mathbf{E} + 2i\omega(\bar{w}_m - \bar{w}_e) \tag{63}$$

can be used as a point of departure. The real part of this equation when integrated throughout a volume V bounded by a closed surface A, which completely encloses the volume V_0 occupied by the current $\mathbf{J}$, yields

$$\text{Re} \int_V \nabla \cdot \mathbf{S} \, dV = -\tfrac{1}{2} \, \text{Re} \int_{V_0} \mathbf{J}^* \cdot \mathbf{E} \, dV \tag{64}$$

Converting the left side by Gauss' theorem to a surface integral over the closed surface A with unit outward normal $\mathbf{n}$, we get

$$\text{Re} \int_A \mathbf{n} \cdot \mathbf{S} \, dA = -\tfrac{1}{2} \, \text{Re} \int_{V_0} \mathbf{J}^* \cdot \mathbf{E} \, dV \tag{65}$$

The right side gives the net time-average power available for radiation and the left side the time-average radiated power crossing A in an outward direction. In agreement with the conservation of power this relation is valid regardless of the size and shape of the closed surface A as long as it completely encloses V_0. Thus we see that the time-average radiated power can be computed by integrating $-(\tfrac{1}{2}) \, \text{Re} \, (\mathbf{J}^* \cdot \mathbf{E})$ throughout V_0 or, alternatively, by integrating $\text{Re} \, (\mathbf{n} \cdot \mathbf{S})$ over any closed surface A enclosing V_0. In one extreme case, A coincides with the boundary A_0 of V_0; in the other, A coincides with the sphere at infinity, A_∞.

The imaginary part of Eq. (63) when integrated throughout V yields

$$\text{Im} \int_A \mathbf{n} \cdot \mathbf{S} \, dA = -\tfrac{1}{2} \, \text{Im} \int_{V_0} \mathbf{J}^* \cdot \mathbf{E} \, dV + 2\omega \int_V (\bar{w}_m - \bar{w}_e) dV \tag{66}$$

As before, A is an arbitrary surface completely enclosing V_0. When A coincides with A_0 this equation becomes

$$\text{Im} \int_{A_0} \mathbf{n} \cdot \mathbf{S} \, dA = -\tfrac{1}{2} \, \text{Im} \int_{V_0} \mathbf{J}^* \cdot \mathbf{E} \, dV + 2\omega(\bar{W}_m^{\text{int}} - \bar{W}_e^{\text{int}}) \tag{67}$$

where

$$\bar{W}_m^{\text{int}} = \int_{V_0} \bar{w}_m \, dV \qquad \bar{W}_e^{\text{int}} = \int_{V_0} \bar{w}_e \, dV \tag{68}$$

denote the (internal) time-average magnetic and electric energies stored inside V_0. When A coincides with A_∞ it becomes

$$\operatorname{Im} \int_{A_\infty} \mathbf{n} \cdot \mathbf{S}\, dA = -\tfrac{1}{2} \operatorname{Im} \int_{V_0} \mathbf{J}^* \cdot \mathbf{E}\, dV + 2\omega(\bar{W}_m^{\text{int}} - \bar{W}_e^{\text{int}}) + 2\omega(\bar{W}_m^{\text{ex}} - \bar{W}_e^{\text{ex}}) \quad (69)$$

where

$$\bar{W}_m^{\text{ex}} = \int_{V-V_0} \bar{w}_m\, dV \qquad \bar{W}_e^{\text{ex}} = \int_{V-V_0} \bar{w}_e\, dV \quad (70)$$

denote the (external) time-average magnetic and electric energies stored outside V_0. In the far zone, $\mathbf{S}$ is purely real and consequently Eq. (69) reduces to

$$2\omega[(\bar{W}_e^{\text{ex}} + \bar{W}_e^{\text{int}}) - (\bar{W}_m^{\text{ex}} + \bar{W}_m^{\text{int}})] = -\tfrac{1}{2} \operatorname{Im} \int_{V_0} \mathbf{J}^* \cdot \mathbf{E}\, dV \quad (71)$$

From this relation it is seen that the volume integral of $-(\tfrac{1}{2}) \operatorname{Im} (\mathbf{J}^* \cdot \mathbf{E})$ throughout V_0 gives 2ω times the difference between the time-average electric and magnetic energies stored in all space, i.e., inside V_0 and outside V_0. A relation involving only the external energies is obtained by subtracting Eq. (71) from Eq. (67), viz.,

$$\operatorname{Im} \int_{A_0} \mathbf{n} \cdot \mathbf{S}\, dA = 2\omega(\bar{W}_e^{\text{ex}} - \bar{W}_m^{\text{ex}}) \quad (72)$$

Now, in accord with the left side of relation (65), we shall find the time-average radiated power by integrating $\operatorname{Re} (\mathbf{n} \cdot \mathbf{S})$ over the sphere at infinity. As was shown in Secs. 2.1 and 2.2, the electric field $\mathbf{E}$ produced by a monochromatic current $\mathbf{J}$ is given by

$$\mathbf{E}(\mathbf{r}) = i\omega\mu \int_{V_0} \mathbf{\Gamma}(\mathbf{r},\mathbf{r}') \cdot \mathbf{J}(\mathbf{r}')dV' \quad (73)$$

where

$$\mathbf{\Gamma}(\mathbf{r},\mathbf{r}') = \left(\mathbf{u} + \frac{1}{k^2} \nabla\nabla\right) \frac{e^{ik|\mathbf{r}-\mathbf{r}'|}}{4\pi|\mathbf{r} - \mathbf{r}'|} \quad (74)$$

Since

$$\nabla \frac{e^{ik|\mathbf{r}-\mathbf{r}'|}}{4\pi|\mathbf{r}-\mathbf{r}'|} = -\nabla' \frac{e^{ik|\mathbf{r}-\mathbf{r}'|}}{4\pi|\mathbf{r}-\mathbf{r}'|} \tag{75}$$

we may write expression (74) in the form

$$\mathbf{\Gamma}(\mathbf{r},\mathbf{r}') = \left(\mathbf{u} + \frac{1}{k^2}\nabla'\nabla'\right)\frac{e^{ik|\mathbf{r}-\mathbf{r}'|}}{4\pi|\mathbf{r}-\mathbf{r}'|} \tag{76}$$

with the double gradient operating with respect to the primed coordinates only. In the far zone, which is defined by

$$r \gg r' \qquad \text{and} \qquad kr \gg 1 \tag{77}$$

where $r = \sqrt{\mathbf{r}\cdot\mathbf{r}}$ and $r' = \sqrt{\mathbf{r}'\cdot\mathbf{r}'}$, the following approximation is valid:

$$|\mathbf{r}-\mathbf{r}'| \equiv \sqrt{r^2 + r'^2 - 2\mathbf{r}\cdot\mathbf{r}'} \approx r - \mathbf{e}_r\cdot\mathbf{r}' \tag{78}$$

where $\mathbf{e}_r(= \mathbf{r}/r)$ is the unit vector in the direction of $\mathbf{r}$. In this approximation we may replace $\exp(ik|\mathbf{r}-\mathbf{r}'|)$ by $\exp[ik(r - \mathbf{e}_r\cdot\mathbf{r}')]$ and $1/|\mathbf{r}-\mathbf{r}'|$ by $1/r$. Accordingly Eq. (76) reduces to

$$\mathbf{\Gamma}(\mathbf{r},\mathbf{r}') = \left(\mathbf{u} + \frac{1}{k^2}\nabla'\nabla'\right)\frac{e^{ikr}}{4\pi r}\, e^{-ik\mathbf{e}_r\cdot\mathbf{r}'} \tag{79}$$

in the far zone. The double gradient $\nabla'\nabla'$ operates on $e^{-ik\mathbf{e}_r\cdot\mathbf{r}'}$ only, and since

$$\nabla' e^{-ik\mathbf{e}_r\cdot\mathbf{r}'} = -ik\mathbf{e}_r e^{-ik\mathbf{e}_r\cdot\mathbf{r}'} \tag{80}$$

we have

$$\nabla'\nabla' e^{-ik\mathbf{e}_r\cdot\mathbf{r}'} = -k^2\mathbf{e}_r\mathbf{e}_r e^{-ik\mathbf{e}_r\cdot\mathbf{r}'} \tag{81}$$

With the aid of this relation, expression (79) for the far zone $\mathbf{\Gamma}$ becomes

$$\mathbf{\Gamma}(\mathbf{r},\mathbf{r}') = (\mathbf{u} - \mathbf{e}_r\mathbf{e}_r)\frac{e^{ikr}}{4\pi r}\, e^{-ik\mathbf{e}_r\cdot\mathbf{r}'} \tag{82}$$

Substituting this expression into Eq. (73) and using the vector identity $(\mathbf{u} - \mathbf{e}_r\mathbf{e}_r) \cdot \mathbf{J} = \mathbf{J} - \mathbf{e}_r(\mathbf{e}_r \cdot \mathbf{J}) = -\mathbf{e}_r \times (\mathbf{e}_r \times \mathbf{J})$ we obtain the following representation for the far-zone electric field:

$$\mathbf{E}(\mathbf{r}) = -i\omega\mu \frac{e^{ikr}}{4\pi r} \mathbf{e}_r \times \left[\mathbf{e}_r \times \int_{V_0} e^{-ik\mathbf{e}_r \cdot \mathbf{r}'} \mathbf{J}(\mathbf{r}') dV' \right] \tag{83}$$

The far-zone magnetic field is found by taking the curl of Eq. (83) in accord with the Maxwell equation $\mathbf{H} = \dfrac{1}{i\omega\mu} \nabla \times \mathbf{E}$. Thus

$$\mathbf{H}(\mathbf{r}) = ik \frac{e^{ikr}}{4\pi r} \left[\mathbf{e}_r \times \int_{V_0} e^{-ik\mathbf{e}_r \cdot \mathbf{r}'} \mathbf{J}(\mathbf{r}') dV' \right] \tag{84}$$

Comparing expressions (83) and (84) we see that the far-zone **E** and **H** are perpendicular to each other and to $\mathbf{e}_r$, in agreement with the fact that any far-zone electromagnetic field is purely transverse to the direction of propagation, viz., in the far zone

$$\mathbf{E} = \sqrt{\frac{\mu}{\epsilon}} (\mathbf{H} \times \mathbf{e}_r) \qquad \text{or} \qquad \mathbf{H} = \sqrt{\frac{\epsilon}{\mu}} (\mathbf{e}_r \times \mathbf{E}) \tag{85}$$

is always valid. Expressions (83) and (84) yield the following expression for the far-zone Poynting vector:

$$\mathbf{S} = \tfrac{1}{2}\mathbf{E} \times \mathbf{H}^* = \mathbf{e}_r \sqrt{\frac{\mu}{\epsilon}} \frac{k^2}{32\pi^2 r^2} \left| \mathbf{e}_r \times \int_{V_0} e^{-ik\mathbf{e}_r \cdot \mathbf{r}'} \mathbf{J}(\mathbf{r}') dV' \right|^2 \tag{86}$$

The notation $|\mathbf{C}|^2$ where **C** is any vector means $\mathbf{C} \cdot \mathbf{C}^*$. From expression (86) we see that **S** is purely real and purely radial, i.e., directed parallel to $\mathbf{e}_r$. The element of area of the sphere over which **S** is to be integrated is $r^2\, d\Omega$, where $d\Omega$ is an element of solid angle. Hence, the time-average radiated power P is given by

$$\begin{aligned} P &= \int_{A_\infty} \mathbf{n} \cdot \mathbf{S}\, dA = \int \mathbf{e}_r \cdot \mathbf{S} r^2\, d\Omega \\ &= \sqrt{\frac{\mu}{\epsilon}} \frac{k^2}{32\pi^2} \int d\Omega \left| \mathbf{e}_r \times \int_{V_0} e^{-ik\mathbf{e}_r \cdot \mathbf{r}'} \mathbf{J}(\mathbf{r}') dV' \right|^2 \end{aligned} \tag{87}$$

This way of calculating the radiated power is called the "Poynting vector method."

A formally different way of calculating the radiated power consists in integrating throughout V_0 the quantity $-(\frac{1}{2})\,\mathrm{Re}\,(\mathbf{J}^* \cdot \mathbf{E})$ in which $\mathbf{E}$ is taken to be the radiative electric field as given by Eq. (73). This alternative procedure, which was proposed by Brillouin,[1] yields

$$P = -\tfrac{1}{2}\,\mathrm{Re}\int_{V_0} \mathbf{J}^* \cdot \mathbf{E}\,dV$$

$$= \frac{\omega\mu}{8\pi}\int_{V_0}\int_{V_0} \mathbf{J}^*(\mathbf{r}) \cdot \left(\mathbf{u} + \frac{1}{k^2}\nabla\nabla\right)\frac{\sin\,(k|\mathbf{r} - \mathbf{r}'|)}{|\mathbf{r} - \mathbf{r}'|} \cdot \mathbf{J}(\mathbf{r}')dV'\,dV \quad (88)$$

and is called the "emf method" since it makes use of the induced electromotive force (emf) of the radiative electric field.

Although representations (87) and (88) of these two methods are apparently different, they nevertheless yield the same result for P and in this sense are consistent. To exemplify this we now apply these two methods to the relatively simple case of a thin straight-wire antenna. The antenna has a length $2l$ and lies along the z axis of a cartesian coordinate system with origin at the center of the wire. Since the wire is thin, the antenna current is closely approximated by the filamentary current

$$\mathbf{J} = \mathbf{e}_z I_0 \delta(x)\delta(y)f(z) \quad (89)$$

where I_0 is the reference current, $\mathbf{e}_z$ is the unit vector in the z direction, and $f(z)$ is generally a complex function of the real variable z. By use of this current we get

$$\int_{V_0} e^{-ik\mathbf{e}_r\cdot\mathbf{r}'}\mathbf{J}(\mathbf{r}')dV' = \mathbf{e}_z I_0 \int_{-l}^{l} e^{-ikz'\cos\theta} f(z')dz' \quad (90)$$

where θ is the colatitude in the spherical coordinate system (r,θ,ϕ) defined by $x = r\sin\theta\cos\phi$, $y = r\sin\theta\sin\phi$, and $z = r\cos\theta$. Denoting the unit vectors in the r, θ, and ϕ directions respectively by $\mathbf{e}_r$, $\mathbf{e}_\theta$, and $\mathbf{e}_\phi$ and noting that $\mathbf{e}_r \times \mathbf{e}_z = -\mathbf{e}_\phi \sin\theta$, we find from Eq. (90) and

[1] L. Brillouin, Origin of Radiation Resistance, *Radioelectricité*, April, 1922.

expression (87) that the Poynting vector method yields

$$P = \sqrt{\frac{\mu}{\epsilon}} \frac{k^2}{16\pi} I_0 I_0^* \int_{-l}^{l} \int_{-l}^{l} f(z') f^*(z) dz\, dz' \int_0^\pi e^{ik(z-z')\cos\theta} \sin^3\theta\, d\theta \tag{91}$$

Moreover, by substituting the current as given in Eq. (89) into expression (88), we see that the emf method yields

$$P = \frac{\omega\mu}{8\pi} I_0 I_0^* \int_{-l}^{l} \int_{-l}^{l} f(z') f^*(z) \left(1 + \frac{1}{k^2} \frac{\partial^2}{\partial z^2}\right) \frac{\sin(k|z-z'|)}{|z-z'|} dz\, dz' \tag{92}$$

To show that expressions (91) and (92) are equivalent, we invoke the following elementary results:

$$\int_0^\pi e^{ik(z-z')\cos\theta} \sin^3\theta\, d\theta = \frac{4}{u^2}\left(\frac{\sin u}{u} - \cos u\right)$$

$$\left(1 + \frac{1}{k^2}\frac{\partial^2}{\partial z^2}\right) \frac{\sin(k|z-z'|)}{|z-z'|} = \frac{2k}{u^2}\left(\frac{\sin u}{u} - \cos u\right)$$

where $u = k(z - z')$. With the aid of these results and the introduction of the new variables $\xi = kz$ and $\eta = kz'$ expressions (91) and (92) pass into the common form

$$P = \frac{1}{4\pi}\sqrt{\frac{\mu}{\epsilon}} I_0 I_0^* \int_{-kl}^{kl} \int_{-kl}^{kl} d\eta\, d\xi \frac{f(\eta) f^*(\xi)}{(\xi-\eta)^2} \left[\frac{\sin(\xi-\eta)}{\xi-\eta} - \cos(\xi-\eta)\right] \tag{93}$$

Thus we see that the Poynting vector method and the emf method ultimately lead to the same formula (93) for the time-average radiated power P and hence are consistent with each other.

From a practical viewpoint, formula (93) as it stands is too clumsy to use, owing to the presence of the double integral. However, Bouwkamp by successive transformations succeeded in reducing the double integral to a repeated integral and then finally to an elegant form involving only single integrals. To demonstrate the capabilities of this form he applied it to several "classical" cases which had been handled previously by the Poynting vector method. For details we refer the reader to his original paper.[1]

The present discussion may be extended to the case of magnetic currents by using the duality transformations of Sec. 1.2. For example, if we replace $\mathbf{J}$, $\mathbf{E}$, $\mathbf{H}$ respectively by $-\sqrt{\epsilon/\mu}\,\mathbf{J}_m$, $-\sqrt{\mu/\epsilon}\,\mathbf{H}$, $\sqrt{\epsilon/\mu}\,\mathbf{E}$ in the far-zone field formulas (83) and (84), we obtain the corresponding formulas for the far-zone field of a monochromatic magnetic current density:

$$\mathbf{H} = -i\omega\epsilon \frac{e^{ikr}}{4\pi r} \mathbf{e}_r \times \left(\mathbf{e}_r \times \int_{V_0} e^{-ik\mathbf{e}_r \cdot \mathbf{r}'} \mathbf{J}_m(\mathbf{r}') dV' \right) \tag{94}$$

$$\mathbf{E} = -ik \frac{e^{ikr}}{4\pi r} \left(\mathbf{e}_r \times \int_{V_0} e^{-ik\mathbf{e}_r \cdot \mathbf{r}'} \mathbf{J}_m(\mathbf{r}') dV' \right) \tag{95}$$

The Poynting vector of this far-zone electromagnetic field is

$$\mathbf{S} = \tfrac{1}{2}\mathbf{E} \times \mathbf{H}^* = \mathbf{e}_r \sqrt{\frac{\epsilon}{\mu}} \frac{k^2}{32\pi^2 r^2} \left| \mathbf{e}_r \times \int_{V_0} e^{-ik\mathbf{e}_r \cdot \mathbf{r}'} \mathbf{J}_m(\mathbf{r}') dV' \right|^2 \tag{96}$$

and consequently the time-average radiated power is

$$P = \int_{A_\infty} \mathbf{n} \cdot \mathbf{S}\, dA = \int_{A_\infty} \mathbf{e}_r \cdot \mathbf{S} r^2\, d\Omega$$

$$= \sqrt{\frac{\epsilon}{\mu}} \frac{k^2}{32\pi^2} \int d\Omega \left| \mathbf{e}_r \times \int_{V_0} e^{-ik\mathbf{e}_r \cdot \mathbf{r}'} \mathbf{J}_m(\mathbf{r}') dV' \right|^2 \tag{97}$$

[1] C. J. Bouwkamp, *Philips Res. Rept.*, **1:** 65 (1946).

3 Radiation from wire antennas

As a practical source of monochromatic radiation the wire antenna plays an important role. The field radiated by such an antenna can be obtained from a knowledge of its current distribution by using the formulas derived in the previous chapter. Although the determination of the antenna current is a boundary-value problem of considerable complexity, a sufficiently accurate estimate of the current distribution can be obtained in the case of thin wires by assuming that the antenna current is a solution of the one-dimensional Helmholtz equation and hence consists of an appropriate superposition of simple waves of current. This simplifying approximation yields satisfactory results for the far-zone field and for those quantities that depend on the far-zone field, e.g., radiation resistance and gain, because the far-zone field in almost all directions is insensitive to small deviations of the current from the exact current. The radiation properties of thin-wire antennas and their arrays are discussed in this chapter.

3.1 Simple Waves of Current

We consider a straight-wire antenna lying along the z axis of a cartesian coordinate system with one end at $z = -l$ and the other at $z = l$, as shown in Fig. 3.1. Since the wire is

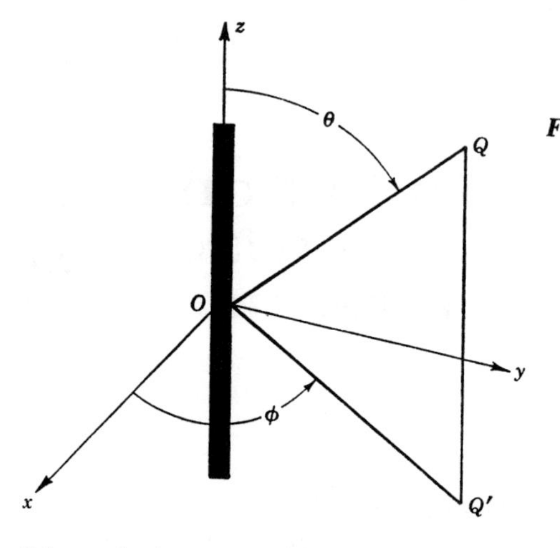

Fig. 3.1 ***Coordinate system for a straight-wire antenna extending from $z = -l$ to $z = l$. Q is observation point. Q' is projection of Q in x-y plane.***

thin and since we wish to calculate only the far-zone field it is a permissible mathematical idealization to assume that the antenna current density is the filamentary distribution

$$\mathbf{J} = \mathbf{e}_z \delta(x)\delta(y)f(z) \tag{1}$$

The total current is the integral of this distribution over the cross section of the wire:

$$\mathbf{I}(z) = \mathbf{e}_z I(z) = \int \mathbf{J}\, dx\, dy = \mathbf{e}_z f(z) \int \delta(x)\delta(y) dx\, dy = \mathbf{e}_z f(z) \tag{2}$$

It is supposed that the wire is cut at some cross section $z = \eta$ and a monochromatic emf is applied across the gap. The current is necessarily a continuous function of z, but the z derivative of the current may be discontinuous at the gap. The antenna is said to be "center-fed" when $\eta = 0$ and "asymmetrically fed" when $\eta \neq 0$.

For a center-fed antenna, $f(z)$ is a symmetrical function of z and satisfies the one-dimensional Helmholtz equation[1]

$$\frac{d^2 f}{dz^2} + k^2 f = 0 \qquad k = \omega/c = 2\pi/\lambda \tag{3}$$

as well as the end conditions

$$f(l) = f(-l) = 0 \tag{4}$$

[1] It appears that Pocklington was the first to show that the currents along straight or curved thin wires in a first approximation satisfy the Helmholtz equation. See H. C. Pocklington, *Proc. Cambridge Phil. Soc.*, **9:** 324 (1897).

The two independent solutions of Eq. (3) are the simple waves e^{ikz} and e^{-ikz}. Accordingly a general form for the current is

$$I(z) = Ae^{ikz} + Be^{-ikz} \tag{5}$$

where A and B are constants. Writing this form for the two segments of the antennas, we have

$$\begin{aligned} I_1(z) &= A_1e^{ikz} + B_1e^{-ikz} \qquad \text{for } 0 \leq z \leq l \\ I_2(z) &= A_2e^{ikz} + B_2e^{-ikz} \qquad \text{for } -l \leq z \leq 0 \end{aligned} \tag{6}$$

When applied to these expressions, the end conditions (4) yield

$$A_1e^{ikl} + B_1e^{-ikl} = 0 \qquad A_2e^{-ikl} + B_2e^{ikl} = 0 \tag{7}$$

from which it follows that $B_1/A_1 = -e^{2ikl}$, $B_2/A_2 = -e^{-2ikl}$. With these results, Eqs. (6) become

$$\begin{aligned} I_1(z) &= -2iA_1e^{ikl} \sin k(l - z) \qquad \text{for } 0 \leq z \leq l \\ I_2(z) &= 2iA_2e^{-ikl} \sin k(l + z) \qquad \text{for } -l \leq z \leq 0 \end{aligned} \tag{8}$$

The continuity condition $I_1(0) = I_2(0)$ requires that A_1 and A_2 be related by

$$A_1e^{ikl} = -A_2e^{-ikl} \tag{9}$$

In view of this connection between A_1 and A_2 it follows from Eqs. (8) that the current distribution, apart from an arbitrary multiplicative constant I_0, is given by the standing wave[1]

$$I(z) = I_0 \sin k(l - |z|) \tag{10}$$

which, for several typical cases, is displayed in Fig. 3.2. This "sinusoidal approximation" is adequate for the purpose of computing the far-zone radiation pattern of a center-fed straight-wire antenna, provided the antenna is neither "too thick" nor "too long." A closer approximation to the true current may be obtained heuristically by adding to the sinusoidal current a quadrature current, which takes into

[1] J. Labus, *Z. Hochfrequenztechnik und Elektrokustik*, **41:** 17 (1933).

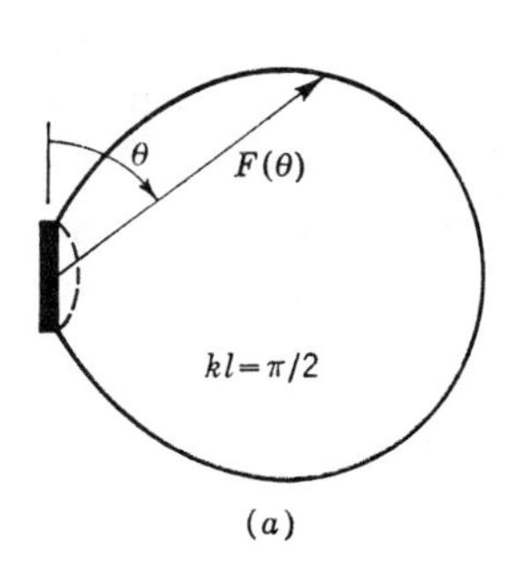

(a)

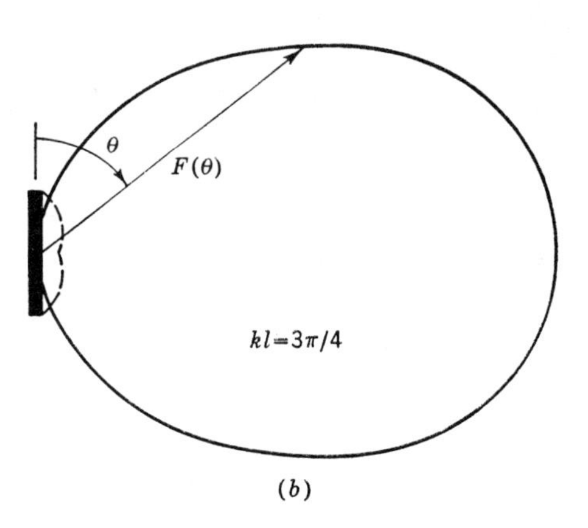

(b)

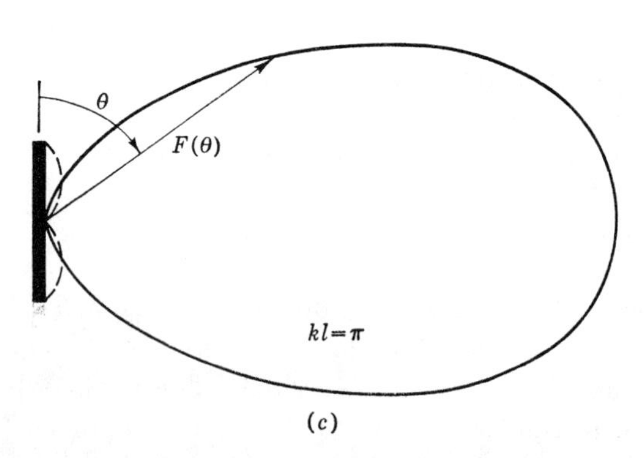

(c)

Fig. 3.2 ***Radiation patterns of a center-driven thin-wire antenna of current distribution shown by dotted lines.***

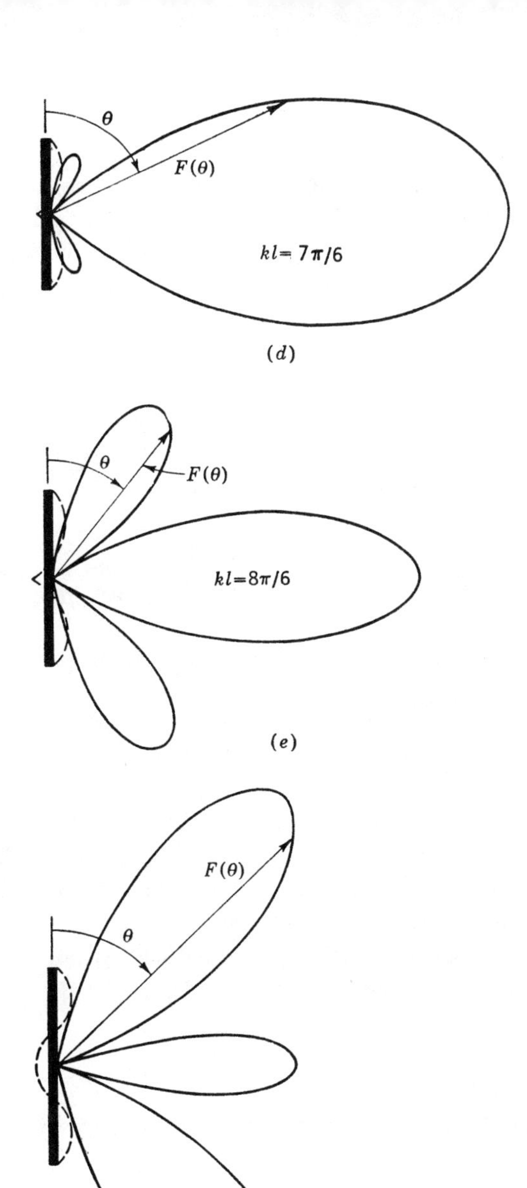

various lengths shown by solid lines. Assumed sinusoidal

account the reaction of the radiation and the ohmic losses on the current.[1] Since the antenna is center-fed, the principal alteration that such a quadrature current can make on the far-zone radiation pattern is the presently negligible one of relaxing the intermediate nulls of the pattern.[2]

However, for an asymmetrically fed antenna ($\eta \neq 0$) the radiation pattern calculated solely on the basis of a simple standing wave of current can be in serious error due to the presence of traveling waves of current. The standing wave, which for arbitrary values of η has the form[3]

$$\begin{aligned} I(z) &= I_0 \sin k(l + \eta) \sin k(l - z) \qquad \text{for } \eta \leq z \leq l \\ I(z) &= I_0 \sin k(l - \eta) \sin k(l + z) \qquad \text{for } -l \leq z \leq \eta \end{aligned} \tag{11}$$

always gives rise to a radiation pattern that is symmetrical about the plane $\theta = \pi/2$. On the other hand, a traveling wave produces an asymmetrical radiation pattern, viz., a pattern tilted toward the direction of the traveling wave. Accordingly the traveling waves tend to tilt the lobes of the pattern and to change their size. Thus if the traveling waves are appreciable, marked changes in the shape of the radiation pattern can occur. Generally the problem of finding the radiation pattern of an asymmetrically driven antenna cannot be handled adequately within the framework of the simple wave theory, except in those cases where either the standing wave or the traveling waves dominate the pattern.

3.2 Radiation from Center-driven Antennas

As indicated by Eqs. (83), (84), and (86) of Chap. 2, the calculation of the far-zone radiation emitted by a distribution of monochromatic

[1] Ronold King and C. W. Harrison, Jr., *Proc. IRE*, **31:** 548 (1943).

[2] C. W. Harrison, Jr., and Ronold King, *Proc. IRE*, **31:** 693 (1943).

[3] See, for example, S. A. Schelkunoff and H. T. Friis, "Antennas: Theory and Practice," chap. 8, John Wiley and Sons, Inc., New York, 1952.

current centers on the evaluation of the so-called radiation vector **N** defined by the integral[1]

$$\mathbf{N} = \int_{V_0} e^{-ik\mathbf{e}_r\cdot\mathbf{r}'}\mathbf{J}(\mathbf{r}')dV' \tag{12}$$

where $\mathbf{e}_r$ is the unit vector pointing from the origin to the point of observation and $\mathbf{r}'$ is the position vector extending from the origin to the volume element dV'. The required information on **J** can be obtained either by solving the boundary-value problem which the analytical determination of **J** poses or by choosing the current on empirical grounds. In the present case of a thin-wire antenna, the latter alternative is adopted, according to which it is alleged that a sufficiently accurate representation of the antenna current can be built from simple waves to agree with the results of measurement.

Accordingly, let us consider the case of a center-driven thin-wire antenna lying along the z axis with one end at $z = -l$ and the other at $z = l$. It is known a posteriori that the current distribution along such an antenna may be approximated, insofar as the far-zone radiation is concerned, by the sinusoidal filamentary current

$$\mathbf{J}(\mathbf{r}) = \mathbf{e}_z I_0 \delta(x)\delta(y) \sin k(l - |z|) \tag{13}$$

Substituting this assumed current into definition (12) and performing the integrations with respect to x' and y', we get the one-dimensional integral

$$\mathbf{N} = \mathbf{e}_z I_0 \int_{-l}^{l} e^{-ikz'\cos\theta} \sin k(l - |z'|)dz' \tag{14}$$

which by use of the integration formula

$$\int e^{a\xi} \sin (b\xi + c)d\xi = \frac{e^{a\xi}}{a^2 + b^2}[a \sin (b\xi + c) - b \cos (b\xi + c)] \tag{15}$$

yields

$$\mathbf{N} = \mathbf{e}_z 2I_0 \frac{\cos (kl \cos \theta) - \cos kl}{k \sin^2 \theta} \tag{16}$$

[1] S. A. Schelkunoff, A General Radiation Formula, *Proc. IRE*, **27:** 660–666 (1939).

With the aid of this result and the vector relations $\mathbf{e}_r \times \mathbf{e}_z = -\mathbf{e}_\phi \sin\theta$, $\mathbf{e}_r \times (\mathbf{e}_r \times \mathbf{e}_z) = \mathbf{e}_\theta \sin\theta$, it follows from Eqs. (83), (84), and (86) of Chap. 2 that the far-zone electric and magnetic fields are

$$E_\theta = -i\sqrt{\frac{\mu}{\epsilon}}\frac{e^{ikr}}{2\pi r} I_0 \frac{\cos(kl\cos\theta) - \cos kl}{\sin\theta} \tag{17}$$

and

$$H_\phi = -i\frac{e^{ikr}}{2\pi r} I_0 \frac{\cos(kl\cos\theta) - \cos kl}{\sin\theta} \tag{18}$$

and that the radial component of the Poynting vector is

$$S_r = \sqrt{\frac{\mu}{\epsilon}}\frac{I_0^2}{8\pi^2 r^2}\left[\frac{\cos(kl\cos\theta) - \cos kl}{\sin\theta}\right]^2 \tag{19}$$

In these expressions, the common factor

$$F(\theta) = \frac{\cos(kl\cos\theta) - \cos kl}{\sin\theta} \tag{20}$$

is the radiation pattern of the antenna. Since the radiation pattern is independent of ϕ it is said to be "omnidirectional." When the antenna is short compared to the wavelength ($kl \ll 1$) the radiation pattern reduces to[1]

$$F(\theta) = \tfrac{1}{2}(kl)^2 \sin\theta \tag{21}$$

From this we see that the radiation pattern of a short wire antenna consists of a single lobe that straddles the equatorial plane $\theta = \pi/2$ and exhibits nulls at the poles $\theta = 0$ and $\theta = \pi$. As kl increases up to $kl = \pi$ the lobe becomes narrower and more directive. As kl exceeds $kl = \pi$ and approaches $kl = 3\pi/2$, two side lobes appear, gradually growing in size and ultimately becoming larger than the central lobe itself. (See Fig. 3.2.)

Since $F(\theta)$ is an even function of $\theta - \pi/2$ that vanishes at $\theta = 0$ and

[1] In the case of a Hertzian dipole $F(\theta) = kl\sin\theta$. To show this, we recall that the current density of a Hertzian dipole of length $2l$, located at the origin of coordinates and directed parallel to the z axis, is defined as $\mathbf{J} = \mathbf{e}_z I_0 \delta(x)\delta(y)$, then note that for this current $\mathbf{N} = \mathbf{e}_z 2lI_0$ and hence $S_r = \sqrt{\mu/\epsilon}\,(I_0^2/8\pi^2 r^2)(kl\sin\theta)^2$.

$\theta = \pi$, it may be expanded[1] in a Fourier series of the form

$$F(\theta) = \sum_{n=0}^{\infty} b_{2n+1} \sin (2n+1)\theta \qquad (0 \leq \theta \leq \pi) \tag{22}$$

where

$$b_{2n+1} = \frac{2}{\pi} \int_0^\pi F(\theta) \sin (2n+1)\theta \, d\theta \tag{23}$$

For small kl, all the higher-order coefficients are, to a good approximation, negligible compared to the first coefficient

$$b_1 = \frac{2}{\pi} \int_0^\pi [\cos (kl \cos \theta) - \cos kl] d\theta = 2J_0(kl) - 2 \cos kl \tag{24}$$

Although this simple approximation deteriorates as the length of the antenna increases, for a half-wave dipole ($kl = \pi/2$) it is still satisfactory and yields

$$F(\theta) = \frac{\cos \left(\frac{\pi}{2} \cos \theta\right)}{\sin \theta} \approx 0.945 \sin \theta \tag{25}$$

From a practical viewpoint this approximate representation of the distant field of a half-wave dipole provides a useful simplification. For example, it enables one to obtain a convenient expression for the radiation resistance of certain linear arrays of half-wave dipoles.[2]

3.3 Radiation Due to Traveling Waves of Current, Cerenkov Radiation

In the previous section we noted that the far-zone radiation field of a center-fed thin-wire antenna is determined with sufficient accuracy by

[1] R. King, The Approximate Representation of the Distant Field of Linear Radiators, *Proc. IRE,* **29:** 458–463 (1941); C. J. Bouwkamp, On the Effective Length of a Linear Transmitting Antenna, *Philips Res. Rept.,* **4:** 179–188 (1949).

[2] C. H. Papas and Ronold King, The Radiation Resistance of End-fire and Collinear Arrays, *Proc. IRE,* **36:** 736–741 (1948).

using the standing-wave part of the antenna current and ignoring the traveling-wave part. In the present section we shall discuss the converse state of the antenna, wherein the traveling-wave part of the current is dominant and the standing-wave part is quite negligible. Such a state can be achieved by the proper excitation and termination of the antenna.[1]

Accordingly we assume that the current distribution along a thin-wire antenna is the traveling wave

$$\mathbf{J}(\mathbf{r}) = \mathbf{e}_z I_0 \delta(x)\delta(y) e^{ipkz} \qquad (-l \leq z \leq l) \tag{26}$$

Here the index p is the ratio of the velocity of light to the velocity of the current wave along the antenna. This index, which is equal to or greater than unity, depends on the degree to which the antenna is loaded. If the antenna is unloaded, i.e., if the antenna wire is bare, p is approximately equal to unity. Then as the loading is increased[2] there is a corresponding increase in p.

Substituting expression (26) into definition (12) we get

$$\mathbf{N} = \mathbf{e}_z I_0 \int_{-l}^{l} e^{-ikz\cos\theta} e^{ipkz}\, dz = \mathbf{e}_z 2I_0 \frac{\sin\,[kl(p - \cos\theta)]}{k(p - \cos\theta)} \tag{27}$$

This expression for the radiation vector, when introduced into Eqs. (83), (84), and (86) of Chap. 2, yields the following nonvanishing components of the far-zone fields and the Poynting vector:

$$E_\theta = \sqrt{\frac{\mu}{\epsilon}}\, H_\phi = -i\sqrt{\frac{\mu}{\epsilon}}\, \frac{e^{ikr}}{2\pi r}\, I_0 \sin\theta\, \frac{\sin\,[kl(p - \cos\theta)]}{p - \cos\theta} \tag{28}$$

$$S_r = \sqrt{\frac{\mu}{\epsilon}}\, \frac{1}{8\pi^2 r^2}\, I_0{}^2 \sin^2\theta\, \frac{\sin^2\,[kl(p - \cos\theta)]}{(p - \cos\theta)^2} \tag{29}$$

From these expressions it follows that the radiation pattern of the

[1] A practical example of such an antenna is the "wave antenna" or "Beverage antenna." See H. H. Beverage, C. W. Rice, and E. W. Kellogg, The Wave Antenna, a New Type of Highly Directive Antenna, *Trans. AIEE*, **42:** 215 (1923).

[2] The loading may take the form of a dielectric coating or a corrugation of the surface.

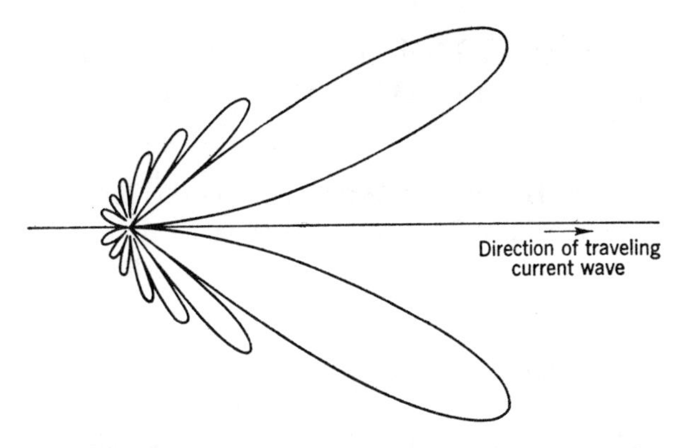

Fig. 3.3 Typical radiation pattern for traveling wave of current.

traveling wave of current (26) is

$$F(\theta) = \sin\theta \frac{\sin [kl(p - \cos\theta)]}{p - \cos\theta} \tag{30}$$

When the antenna is short ($kl \ll 1$), the radiation pattern (30) reduces to

$$F(\theta) = kl \sin\theta \tag{31}$$

Comparing radiation patterns (31) and (21) we see that for short antennas ($kl \ll 1$) the radiation pattern (21) of the standing wave of current (13) has the same form ($\sin\theta$) as the radiation pattern (31) of the traveling wave of current (26).

However, for longer antennas the patterns (30) and (20) differ markedly, the essence of the difference being that the pattern (20) of the standing wave is symmetrical with respect to the equatorial plane $\theta = \pi/2$ whereas the pattern (30) of the traveling wave is asymmetrical. The maxium radiation of the traveling wave appears as a cone in the forward direction, i.e., in the direction of travel of the current wave; the half-angle of the cone decreases as p increases or as kl increases (see Fig. 3.3). This type of conical beam radiation resembles the Cerenkov radiation[1] from fast electrons.

[1] P. A. Cerenkov, *Phys. Rev.*, **52:** 378 (1937). I. Frank and I. Tamm, *Comptes rendus de l'Acad. Sci. U.R.S.S.*, **14:** 109 (1937). See also, J. V. Jelley, "Cerenkov Radiation and Its Applications," Pergamon Press, New York, 1958.

3.4 Integral Relations between Antenna Current and Radiation Pattern

Again we study the thin-wire antenna, but in this instance we do not specify the current distribution. That is, we restrict the current distribution only to the extent of postulating a monochromatic current of the form

$$\mathbf{J}(\mathbf{r}) = \mathbf{e}_z\delta(x)\delta(y)f(z) \qquad (|z| \leq l) \tag{32}$$

where the function $f(z)$ may be complex. From Eq. (83) of Chap. 2, it directly follows that the far-zone electric field of this current distribution is

$$E_\theta = -i\omega\mu \frac{e^{ikr}}{4\pi r} \sin\theta \int_{-l}^{l} e^{-ikz\cos\theta} f(z)dz \qquad (0 \leq \theta \leq \pi) \tag{33}$$

Since the θ-dependent factor of this expression is, by definition, the radiation pattern, we have

$$F(\theta) = \sin\theta \int_{-l}^{l} e^{-ikz\cos\theta} f(z)dz \qquad (0 \leq \theta \leq \pi) \tag{34}$$

This integral relation shows that when $f(z)$ is given in the interval $(|z| \leq l)$, the radiation pattern $F(\theta)$ is uniquely determined for all real angles in the interval $(0 \leq \theta \leq \pi)$.

To proceed toward a relation that would yield $f(z)$ from a knowledge of $F(\theta)$, we cast Eq. (34) into the form of a Fourier integral and then find its mate. Accordingly, the finite limits on the integral in Eq. (34) are replaced with infinite ones by assuming that $f(z)$ vanishes identically outside the interval $(|z| \leq l)$, i.e.,

$$f(z) = 0 \qquad \text{for} \qquad |z| \geq l \tag{35}$$

With $f(z)$ so continued, Eq. (34) can be written as

$$\frac{F(\theta)}{\sin\theta} = \int_{-\infty}^{\infty} e^{-ikz\cos\theta} f(z)dz \qquad (0 \leq \theta \leq \pi) \tag{36}$$

or, in terms of the new variable $\eta(= k \cos\theta)$, as

$$\frac{F(\cos^{-1}\eta/k)}{\sqrt{1-\eta^2/k^2}} = \int_{-\infty}^{\infty} e^{-i\eta z} f(z)dz \qquad (-k \leq \eta \leq k) \tag{37}$$

Now the range of validity of Eq. (37) is extended from $(-k \leq \eta \leq k)$ to $(-\infty \leq \eta \leq \infty)$ by letting θ trace the contour C in the complex θ plane (Fig. 3.4). Such an extension of Eq. (37) leads to the Fourier integral

$$\frac{F(\cos^{-1}\eta/k)}{\sqrt{1-\eta^2/k^2}} = \int_{-\infty}^{\infty} e^{-i\eta z} f(z)dz \qquad (-\infty \leq \eta \leq \infty) \tag{38}$$

By the Fourier integral theorem, the mate of Eq. (38) is

$$f(z) = \frac{1}{2\pi}\int_{-\infty}^{\infty} \frac{F(\cos^{-1}\eta/k)}{\sqrt{1-\eta^2/k^2}} e^{i\eta z}\, d\eta \qquad (-\infty \leq z \leq \infty) \tag{39}$$

Transforming to the complex θ plane (by use of $\eta = k\cos\theta$) and explicitly taking into account the requirement (35) that $f(z)$ vanish for

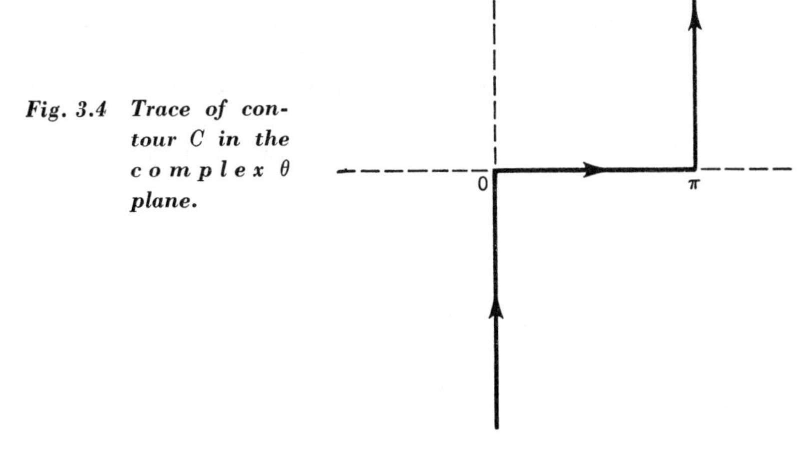

Fig. 3.4 Trace of contour C in the complex θ plane.

$|z| \geq l$, we obtain the desired relation

$$f(z) = \frac{k}{2\pi} \int_C F(\theta) e^{ikz \cos \theta} \, d\theta \qquad (|z| \leq l) \tag{40}$$

and the side condition

$$0 = \frac{k}{2\pi} \int_C F(\theta) e^{ikz \cos \theta} \, d\theta \qquad (|z| \geq l) \tag{41}$$

From Eq. (40) it is clear that $F(\theta)$ must be known along the entire contour C before $f(z)$ can be evaluated from it. Moreover, since $F(\theta)$ must satisfy the side condition (41), it cannot be chosen arbitrarily. Nevertheless, it seems possible[1] to find an $F(\theta)$ which satisfies Eq. (41) and closely approximates a prescribed radiation pattern in the range of real values $(0 \leq \theta \leq \pi)$.

3.5 Pattern Synthesis by Hermite Polynomials

In connection with the antenna of the previous section we now briefly sketch the approximation method of Bouwkamp and De Bruijn,[2] which enables one to calculate a current distribution that will produce a prescribed radiation pattern, or, in other words, enables one to synthesize a given radiation pattern.

The point of departure is the integral relation (34) connecting the radiation pattern $F(\theta)$ to the current distribution $f(z)$. For convenience, however, we express this relation in terms of the dimensionless variables $t = \cos \theta$, $\xi = kz$ and the dimensionless constant $a = kl$. Thus, Eq. (34), apart from an ignorable constant, is written first as

$$F(t) = \sqrt{1 - t^2} \int_{-a}^{a} e^{-it\xi} f(\xi) d\xi \qquad (-1 \leq t \leq 1) \tag{42}$$

[1] For a heuristic discussion of such a possibility, see P. M. Woodward and J. D. Lawson, The Theoretical Precision with which an Arbitrary Radiation Pattern may be obtained from a Source of Finite Size, *J. Inst. Elec. Eng.*, **95** (part III): 363–370 (1948).

[2] C. J. Bouwkamp and N. G. de Bruijn, The Problem of Optimum Antenna Current Distribution, *Philips Res. Rept.*, **1:** 135–138 (1946).

and then, by use of the shorthand

$$G(t) = \frac{F(t)}{\sqrt{1 - t^2}} \tag{43}$$

as

$$G(t) = \int_{-a}^{a} e^{-it\xi} f(\xi) d\xi \qquad (-1 \leq t \leq 1) \tag{44}$$

Referring to this integral equation, we see that the synthesis problem consists in finding $f(\xi)$ when $G(t)$ is given in the interval $-1 \leq t \leq 1$.

By virtue of a theorem due to Weierstrass,[1] we may approximate the given function $G(t)$ by a polynomial $p(t)$ of sufficiently high degree N:

$$G(t) = p(t) \equiv \gamma_0 + \gamma_1 t + \cdots + \gamma_N t^N \tag{45}$$

Moreover, we may invoke unknown functions $f_n(\xi)$ such that

$$f(\xi) = \gamma_0 f_0(\xi) + \gamma_1 f_1(\xi) + \cdots + \gamma_N f_N(\xi) \tag{46}$$

Substituting expressions (45) and (46) into the integral equation (44), we see that functions $f_n(\xi)$ for $n = 0, 1, \ldots, N$ must satisfy

$$t^n = \int_{-a}^{a} e^{-it\xi} f_n(\xi) d\xi \qquad (-1 \leq t \leq 1) \tag{47}$$

To find the functions $f_n(\xi)$, we introduce the Hermite polynomials $H_n(u)$ defined by[2]

$$H_n(u) = (-1)^n e^{u^2/2} \left(\frac{d}{du}\right)^n e^{-u^2/2} \qquad (n = 0, 1, 2, \ldots) \tag{48}$$

From this formula, the following result can be verified by repeated

[1] See, for example, R. Courant and D. Hilbert, "Methods of Mathematical Physics," vol. 1, p. 65, Interscience Publishers, Inc., New York, 1953.

[2] This definition agrees with that of E. T. Whittaker and G. N. Watson, "A Course of Modern Analysis," p. 350, Cambridge University Press, London, 1940. $H_n(u) = e^{u^2/4} D_n(u)$, where $D_n(u)$ is that given by Whittaker and Watson.

partial integrations:

$$\int_{-\infty}^{\infty} e^{-A^2\xi^2/2} H_n(A\xi) e^{-it\xi}\, d\xi = \frac{(-i)^n \sqrt{2\pi}}{A^{n+1}} e^{-t^2/2A^2} t^n \qquad (-1 \leq t \leq 1) \tag{49}$$

When the arbitrary positive constant A is large, the factor exp $(-A^2\xi^2/2)H_n(A\xi)$ decreases rapidly to zero as $|\xi| \to \infty$. Hence, the contribution of the integration beyond a certain range (say, $|\xi| > a$) is negligible. Also, when A is large, the factor exp $(-t^2/2A^2)$ approaches unity. Accordingly, if we choose A sufficiently large, then Eq. (49) closely approximates

$$\frac{A^{n+1}}{2\pi(-i)^n} \int_{-a}^{a} e^{-A^2\xi^2/2} H_n(A\xi) e^{-it\xi}\, d\xi = t^n \qquad (-1 \leq t \leq 1) \tag{50}$$

Comparing Eqs. (50) and (47), we see that the functions $f_n(\xi)$ are given by

$$f_n(\xi) = \frac{A^{n+1}}{\sqrt{2\pi}\,(-i)^n} e^{-A^2\xi^2/2} H_n(A\xi) \tag{51}$$

Substituting this result into Eq. (46) we get the formal solution of integral equation (44):

$$f(\xi) = \frac{1}{\sqrt{2\pi}} \sum_{n=0}^{N} \frac{\gamma_n A^{n+1}}{(-i)^n} e^{-A^2\xi^2/2} H_n(A\xi) \qquad (A \text{ large}) \tag{52}$$

As an application of the above method we now synthesize the radiation pattern

$$F(\theta) = \sin^{2N+1}\theta \tag{53}$$

Since $t = \cos\theta$, then $F(t) = \sqrt{1 - t^2}\,(1 - t^2)^N$ and hence

$$G(t) = (1 - t^2)^N \tag{54}$$

By the binomial theorem, we have

$$G(t) = (1 - t^2)^N = \sum_{l=0}^{N} \binom{N}{l} (-i)^{2l} t^{2l} \tag{55}$$

where the binomial coefficients are given by

$$\binom{N}{l} = \frac{N!}{(N - l)!l!}$$

Comparing expansions (55) and (45) we see that

$$\gamma_{2l} = \binom{N}{l} (-i)^{2l} \tag{56}$$

and substituting these values into Eq. (52) we get

$$f(\xi) = \frac{A}{\sqrt{2\pi}} e^{-A^2\xi^2/2} \sum_{l=0}^{N} \binom{N}{l} A^{2l} H_{2l}(A\xi) \tag{57}$$

The arbitrary positive constant A is chosen such that

$$A = \frac{\beta}{a} \tag{58}$$

where β is greater than the largest root of $H_{2l}(u) = 0$. With the use of expression (58), the current distribution (57) is transformed to

$$f(a,\xi,\beta) = \frac{\beta}{a\sqrt{2\pi}} e^{-\frac{\beta^2}{2}\frac{\xi^2}{a^2}} \sum_{l=0}^{N} \binom{N}{l} \frac{\beta^{2l}}{a^{2l}} H_{2l}\left(\beta \frac{\xi}{a}\right) \tag{59}$$

Thus, corresponding to $N = 0$, $N = 2$, $N = 4$, we have the radiation patterns

$$F = \sin\theta \qquad F = \sin^5\theta \qquad F = \sin^9\theta \tag{60}$$

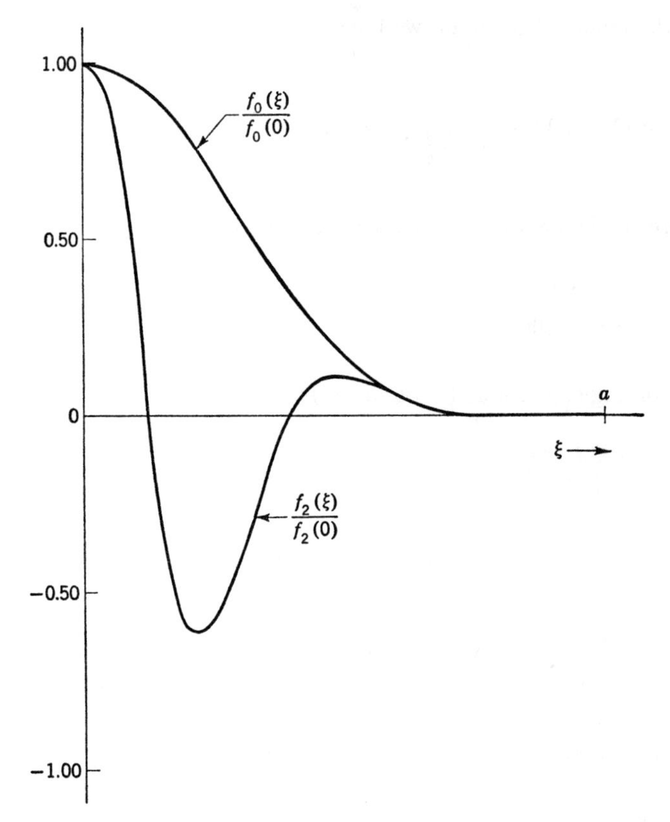

***Fig. 3.5** Current distributions along antenna for $N = 0$ and $N = 2$. Length of antenna is approximately quarter wave.*

and the respective current distributions that produce them:

$$f_0(a,\xi,\beta) = \frac{\beta}{a\sqrt{2\pi}} e^{-\frac{\beta^2}{2}\frac{\xi^2}{a^2}} \tag{61}$$

$$f_2(a,\xi,\beta) = \frac{\beta}{a\sqrt{2\pi}} e^{-\frac{\beta^2}{2}\frac{\xi^2}{a^2}} \left[1 + 2\frac{\beta^2}{a^2}H_2\left(\beta\frac{\xi}{a}\right) + \frac{\beta^4}{a^4}H_4\left(\beta\frac{\xi}{a}\right)\right] \tag{62}$$

$$f_4(a,\xi,\beta) = \frac{\beta}{a\sqrt{2\pi}} e^{-\frac{\beta^2}{2}\frac{\xi^2}{a^2}} \left[1 + 4\frac{\beta^2}{a^2}H_2\left(\beta\frac{\xi}{a}\right)\right.$$
$$\left. + 6\frac{\beta^4}{a^4}H_4\left(\beta\frac{\xi}{a}\right) + 4\frac{\beta^6}{a^6}H_6\left(\beta\frac{\xi}{a}\right) + \frac{\beta^8}{a^8}H_8\left(\beta\frac{\xi}{a}\right)\right] \tag{63}$$

According to the calculations of Bouwkamp and De Bruijn, these distributions are explicitly given by

$$f_0(0.8,\xi,4) = 2.394e^{-12.5\xi^2}$$

$$f_2(\tfrac{3}{4},\xi,6) = 5.175 \times 10^4 e^{-32\xi^2}(1 - 128.663\xi^2 + 1379.59\xi^4)$$

$$f_4(\pi/4,\xi,9) = 1.4208 \times 10^{11} e^{-65\xi^2}(1 - 526\xi^2 + 34{,}559\xi^4 - 605{,}706\xi^6 + 2{,}843{,}678\xi^8)$$

In Figs. 3.5 and 3.6, curves of $f_0(\xi)/f_0(0)$, $f_2(\xi)/f_2(0)$, and $f_4(\xi)/f_4(0)$ versus ξ are plotted. From these curves, we see that as N increases the number of oscillations increases. These spatial oscillations cause

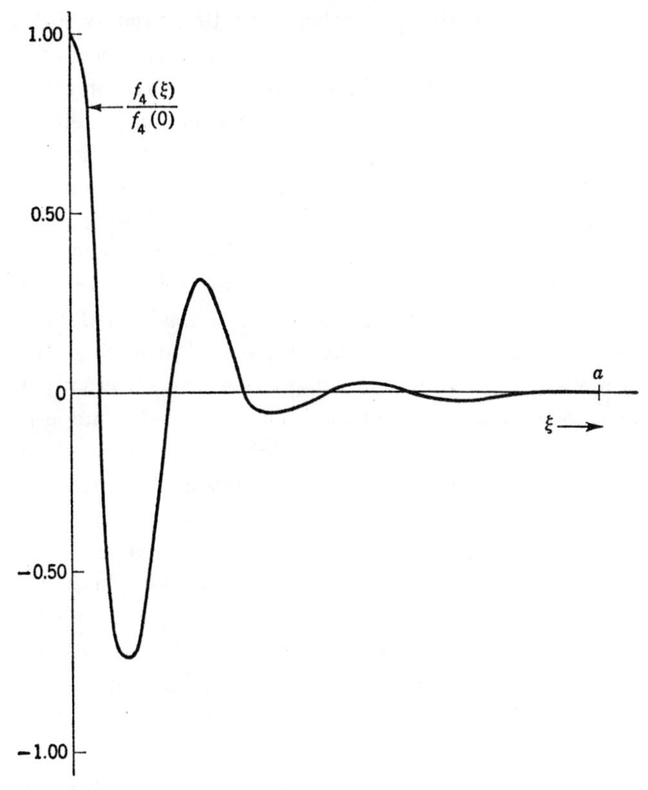

Fig. 3.6 ***Current distribution along antenna for*** $N = 4$. ***Length of antenna is quarter wave.***

the far-zone waves to interfere destructively in every direction except the equatorial one, where they add constructively and thus produce a sharp omnidirectional beam straddling the equatorial plane.

3.6 General Remarks on Linear Arrays

A great variety of radiation patterns can be realized by arranging in space a set of antennas operating at the same frequency. The fields radiated by the separate antennas interfere constructively in certain directions and destructively in others, and thus produce a directional radiation pattern. A knowledge of each antenna's location, orientation, and current distribution, being tantamount to a complete description of the monochromatic source currents, uniquely determines the resultant radiation pattern. Once the vector currents are known, the radiation pattern can be calculated in a straightforward manner by the methods described in Chap. 2. On the other hand, the converse problem of finding a set of antennas that would produce a specified radiation pattern has no unique solution. For this reason, the problem of synthesizing a set of antennas to achieve a prescribed radiation pattern is considerably more challenging than the one of analyzing a prescribed set of antennas for its resultant radiation pattern. Actually, the indeterminacy of the synthesis problem is circumvented by imposing at the start certain constraints on the set which reduce sufficiently its generality and then by specifying the desired radiation pattern with that degree of completeness which would make the problem determinate.[1]

Although any arrangement of antennas can be analyzed for its radiation pattern when the vector current distribution along each of the antennas is known, a synthesis procedure is possible only for certain sets. An important example of such a set is the configuration called the array, which by definition is composed of a finite number of identical antennas, identically oriented, and excited in such a manner that the current distributions on the separate antennas are the same in form but may differ in phase and amplitude. It follows from this definition that

[1] See, for example, Claus Muller, Electromagnetic Radiation Patterns and Sources, *IRE Trans. Antennas Propagation,* **AP-4** (3): 224–232 (1956).

the radiation pattern of an array is always the product of two functions, one representing the radiation pattern of a single antenna in the array and the other, called the array factor or space factor, being interpretable as the radiation pattern of a similar array of nondirective (isotropic) antennas. This separability simplifies the problems of analysis and synthesis to the extent that it permits the actual array to be replaced by a similar array of isotropic antennas.[1]

Of all possible arrays, the linear array is the simplest to handle mathematically and hence constitutes a natural basis for a discussion of antenna arrays. Here we shall limit our attention to linear arrays.[2] Let us consider then a linear array which for definiteness is assumed to consist of n center-driven half-wave dipoles oriented parallel to the z axis with centers at the points $x_p (p = 0, 1, \ldots, n - 1)$ on the x axis (see Fig. 3.7). Each dipole is independently fed, has a length $2l$, and is resonant ($kl = \pi/2$). Under the simplifying approximation that the proximity of the dipoles does not modify the dipole currents or, equivalently, that the dipoles do not interact with each other,[3] the cur-

[1] An isotropic antenna is no more than a conceptual convenience. Actually a system of coherent currents radiating isotropically in all directions of free space is a physical impossibility. This was proved by Mathis using a theorem due to L. E. J. Brouwer concerning continuous vector distributions on surfaces. See H. F. Mathis, A Short Proof that an Isotropic Antenna is Impossible, *Proc. IRE*, **39:** 970 (1951). For another proof see C. J. Bouwkamp and H. B. G. Casimir, On Multipole Expansions in the Theory of Electromagnetic Radiation, *Physica*, **20:** 539 (1954).

[2] For comprehensive accounts of antenna arrays we refer the reader to the excellent literature on the subject. See, for example, G. A. Campbell, "Collected Papers," American Tel. and Tel. Co., New York, 1937; Ronold King, "Theory of Linear Antennas," Harvard University Press, Cambridge, Mass., 1956; S. A. Schelkunoff and H. T. Friis, "Antennas: Theory and Practice," John Wiley & Sons, Inc., New York, 1952; H. Bruckmann, "Antennen ihre Theorie und Technik," S. Hirzel Verlag KG, Stuttgart, 1939; J. D. Kraus, "Antennas," McGraw-Hill Book Company, New York, 1950; H. L. Knudsen, "Bidrag til teorien før antennesystemer med hel eller delvis rotationssymmetri," I Kommission hos Teknick Forlag, Copenhagen, 1952.

[3] In practice, one would take into account this mutual interaction or coupling by invoking the concept of mutual impedance. See, for example, P. S. Carter, Circuit Relations in Radiating Systems and Application to Antenna Problems, *Proc. IRE*, **20:** 1004 (1932); G. H. Brown, Directional Antennas, *Proc. IRE*, **25:** 78 (1937); A. A. Pistolkors, The Radiation Resistance of Beam Antennas, *Proc. IRE*, **17:** 562 (1929); F. H. Murray, Mutual Impedance of Two Skew Antenna Wires, *Proc. IRE*, **21:** 154 (1933).

rent density along the pth dipole is taken to be that of an isolated dipole:

$$\mathbf{J}^{(p)} = \mathbf{e}_z A_p \delta(x - x_p)\delta(y) \cos kz \qquad (-l \leq z \leq l) \tag{64}$$

where A_p denotes the complex magnitude of the current. Hence the resulting current density for the entire array is the sum

$$\mathbf{J} = \sum_{p=0}^{n-1} \mathbf{J}^{(p)} = \mathbf{e}_z \delta(y) \cos kz \sum_{p=0}^{n-1} A_p \delta(x - x_p) \tag{65}$$

This current density gives rise to the following expression for the radia-

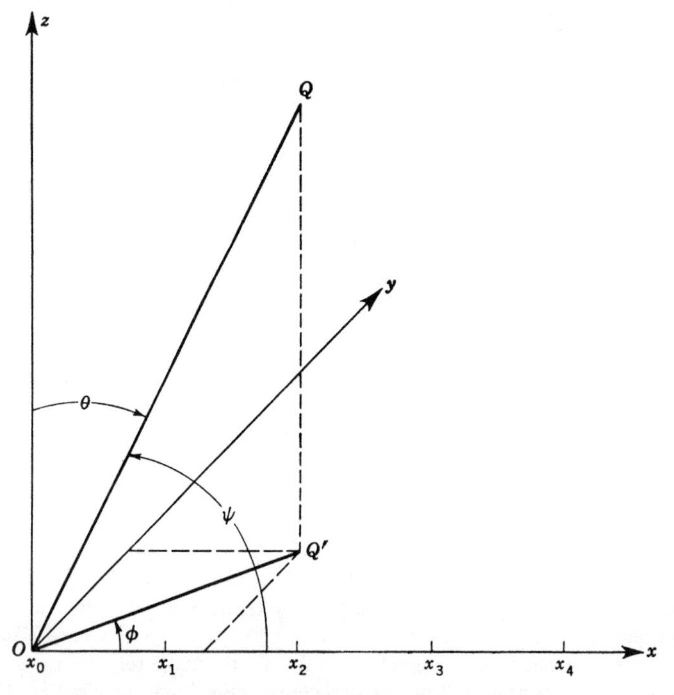

Fig. 3.7 Linear array of half-wave dipoles at points $x_0, x_1, \ldots, x_{n-1}$ along x axis. Each dipole is parallel to z axis. OQ is line of observation. Q' is projection of Q on xy plane.

tion vector:

$$\mathbf{N} = \int e^{-ik\mathbf{e}_r\cdot\mathbf{r}'}\mathbf{J}(\mathbf{r}')dV' = \mathbf{e}_z \int e^{-iky'\sin\theta\sin\phi}\delta(y')dy'$$

$$\times \int_{-l}^{l} e^{-ikz'\cos\theta}\cos kz'\,dz' \int \sum_{p=0}^{n-1} A_p e^{-ikx'\sin\theta\cos\phi}\delta(x' - x_p)dx'$$

which upon integration reduces to

$$\mathbf{N} = \mathbf{e}_z \frac{2}{k} \frac{\cos\left(\frac{\pi}{2}\cos\theta\right)}{\sin^2\theta} \sum_{p=0}^{n-1} A_p e^{-ikx_p\sin\theta\cos\phi} \tag{66}$$

Substituting this result into Eq. (86) of Chap. 2 and recalling the vector relation $\mathbf{e}_r \times \mathbf{e}_z = -\mathbf{e}_\phi \sin\theta$, one finds that the far-zone Poynting vector has only a radial component given by

$$S_r = \sqrt{\frac{\mu}{\epsilon}} \frac{1}{8\pi^2 r^2} |F(\theta)A(\theta,\phi)|^2 \tag{67}$$

where

$$F(\theta) = \frac{\cos\left(\frac{\pi}{2}\cos\theta\right)}{\sin\theta} \tag{68}$$

is the radiation pattern of each dipole, and

$$A(\theta,\phi) = \sum_{p=0}^{n-1} A_p e^{-ikx_p\sin\theta\cos\phi} \tag{69}$$

is the array factor. The radiation pattern of the entire array is

$$U(\theta,\phi) = |F(\theta)A(\theta,\phi)| = F(\theta)|A(\theta,\phi)| \tag{70}$$

If we let ψ denote the angle between the x axis and the line of observation ($\cos\psi = \sin\theta\cos\phi$), the array factor (69) takes the form

$$A(\psi) = \sum_{p=0}^{n-1} A_p e^{-ikx_p\cos\psi} \tag{71}$$

which is recognized as the canonical expression for the complex radiation pattern of a similar array of isotropic antennas. Thus the radiation pattern $U(\theta,\phi)$ of the actual array is equal to the radiation pattern $F(\theta)$ of a dipole multiplied by the radiation pattern $A(\psi)$ of the similar array of isotropic radiators. More generally, expression (71) is valid for any linear array irrespective of the type of its member antennas. For example, if each half-wave dipole of the array were replaced by an antenna having a radiation pattern $G(\theta,\phi)$, then the resulting radiation pattern would be given by $U(\theta,\phi) = |G(\theta,\phi)A(\theta,\phi)|$.

The linear array considered above includes certain special cases which are distinguished by the restrictions one imposes on the complex magnitudes A_p of the input currents and on the positions x_p of the antennas. One such case is that of an equidistantly spaced linear array for which

$$x_p = pd \tag{72}$$

where d is the uniform spacing. Imposing this spatial restriction (72) on the array factor (71) and expressing A_p as the product

$$A_p = a_p e^{-ip\gamma} \tag{73}$$

which explicitly exhibits through the factor $\exp(-ip\gamma)$ the progressive phasing γ of the currents, we get

$$A(\psi) = \sum_{p=0}^{n-1} a_p e^{-ip(kd\cos\psi+\gamma)} = \sum_{p=0}^{n-1} a_p e^{ip\alpha} \tag{74}$$

where the shorthand $\alpha = -kd\cos\psi - \gamma$ has been used. Then, if we introduce the complex variable ξ defined by

$$\xi = e^{i\alpha} \tag{75}$$

the array factor (74) takes the form of a polynomial of degree $n - 1$ in the complex variable ξ:

$$A(\psi) = \sum_{p=0}^{n-1} a_p \xi^p \tag{76}$$

Since the coefficients a_p are arbitrary, some of them may be zero. When this occurs, the antennas which correspond to the vanishing coefficients are absent from the array and the remaining antennas do not necessarily constitute an equidistantly spaced array. Nevertheless, an incomplete array of this sort can be considered equidistantly spaced by regarding d as the "apparent spacing" and n as the "apparent number" of antennas. Thus we see that the polynomial (76) can be identified with any linear array having commensurable separations. The importance of this one-to-one correspondence between polynomial and array stems from the fact that it permits application of the highly developed algebraic theory of polynomials to the synthesis problem. A case in point is Schelkunoff's well-known synthesis procedure,[1] which ingeneously exploits certain algebraic properties of the polynomial (76).

When the coefficients a_p of the polynomial (76) are equal to a constant, which for the present may be taken as unity, the array is said to be "uniform." The array factor of such a uniform linear array with commensurable separations has the closed form

$$A(\psi) = \sum_{p=0}^{n-1} \xi^p = \frac{\xi^n - 1}{\xi - 1} \tag{77}$$

which, with the aid of $\xi = \exp(i\alpha)$, becomes

$$A(\psi) = e^{i(n-1)\alpha/2} \frac{\sin(n\alpha/2)}{\sin(\alpha/2)} \tag{78}$$

Consequently the radiation pattern of a uniform linear array of equidistantly spaced isotropic sources is given by

$$|A(\psi)| = \left| \frac{\sin(n\alpha/2)}{\sin(\alpha/2)} \right| = \left| \frac{\sin[n(kd\cos\psi + \gamma)/2]}{\sin[(kd\cos\psi + \gamma)/2]} \right| \tag{79}$$

It is sometimes convenient to divide $A(\psi)$ by n and thus normalize its

[1] S. A. Schelkunoff, A Mathematical Theory of Linear Arrays, *Bell System Tech. J.*, **22:** 80 (1943).

maximum value to unity. The resulting function $K(\psi)$ is the "normalized radiation characteristic" of the array and is given by

$$K(\psi) = \frac{1}{n} \left| \frac{\sin [n(kd \cos \psi + \gamma)/2]}{\sin [(kd \cos \psi + \gamma)/2]} \right| \tag{80}$$

If the sources are in phase with each other ($\gamma = 0$) and if the spacing is less than a wavelength ($kd < 2\pi$), the radiation characteristic $K(\psi)$ consists of a single major lobe straddling the plane $\psi = \pi/2$ and a number of secondary lobes or "side lobes." As long as the spacing remains less than a wavelength, the spacing has only a secondary effect upon the radiation pattern. Hence, if $\gamma = 0$ and if $kd < 2\pi$, the radiation is cast principally in the broadside direction and the array operates as a "broadside array." However, if the spacing becomes greater than a wavelength ($kd > 2\pi$), the radiation characteristic $K(\psi)$ changes markedly; it develops a multilobe structure consisting of "grating lobes" which collectively resemble the diffraction pattern of a linear optical grating.[1] On the other hand, if the sources are phased progressively such that $kd = -\gamma$ or $kd = \gamma$, the radiation is cast principally in the direction of the line of sources and the array operates as an "end-fire array." If the spacing is less than a half wavelength ($kd < \pi$), there is a single end-fire lobe in the direction $\psi = \pi$ when $kd = \gamma$. But if the spacing is equal to a half wavelength ($kd = \pi$), two end-fire lobes exist simultaneously, one along $\psi = 0$ and the other along $\psi = \pi$. Hence, when $kd < \pi$ the array is a "unilateral end-fire array" and when $kd = \pi$ it is a "bilateral end-fire array." An increase in the directivity of a unilateral end-fire array is realized when the condition of Hansen and Woodyard is satisfied, viz., $\gamma = -(kd + \pi/n)$ or $\gamma = (kd + \pi/n)$.[2] If one desires the major lobe to point in some arbitrary direction $\psi = \psi_1$, then the phase γ and the spacing d must be chosen such that $kd \cos \psi_1 + \gamma = 0$.

When the coefficients a_p of the polynomial (76) are smoothly tapered in accord with the binomial coefficients

$$a_p = \binom{n-1}{p} = \frac{(n-1)!}{(n-1-p)!p!} \tag{81}$$

[1] See, for example, A. Sommerfeld, "Optics," pp. 180–185, Academic Press Inc., New York, 1954.

[2] W. W. Hansen and J. R. Woodyard, A New Principle in Directional Antenna Design, *Proc. IRE*, **26:** 333 (1938).

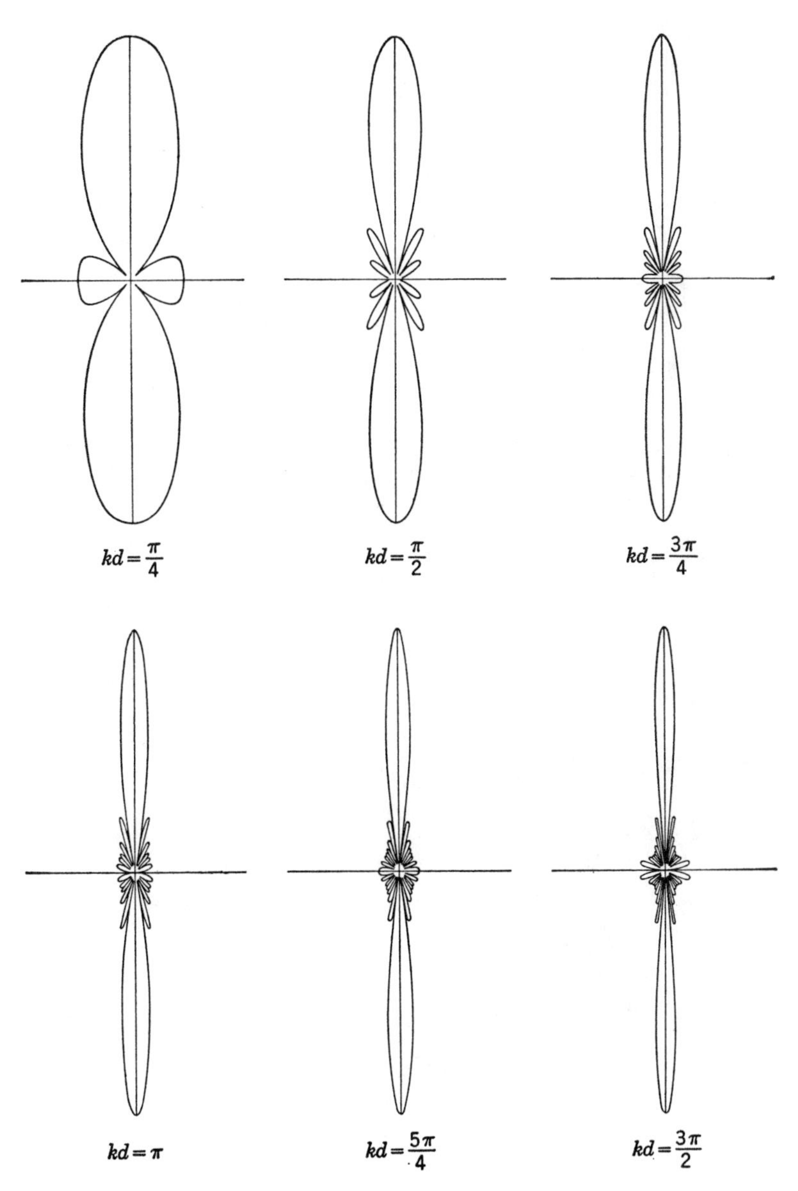

Fig. 3.8 ***Radiation characteristic*** $K(\psi)$ ***of a uniform linear array for various spacings. Calculated from Eq.*** (80) ***with*** $\gamma = 0$ ***and*** $n = 12$. ***Broadside array. Grating lobes.***

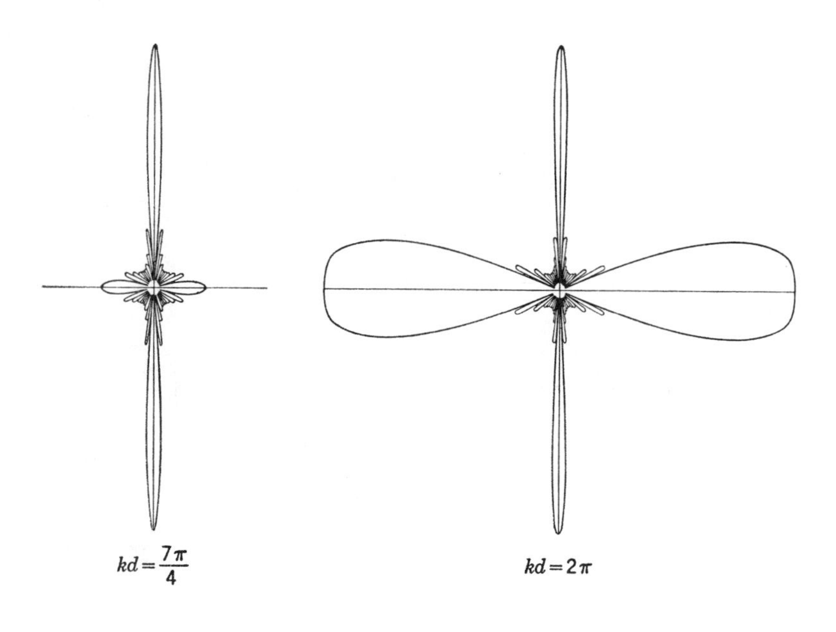

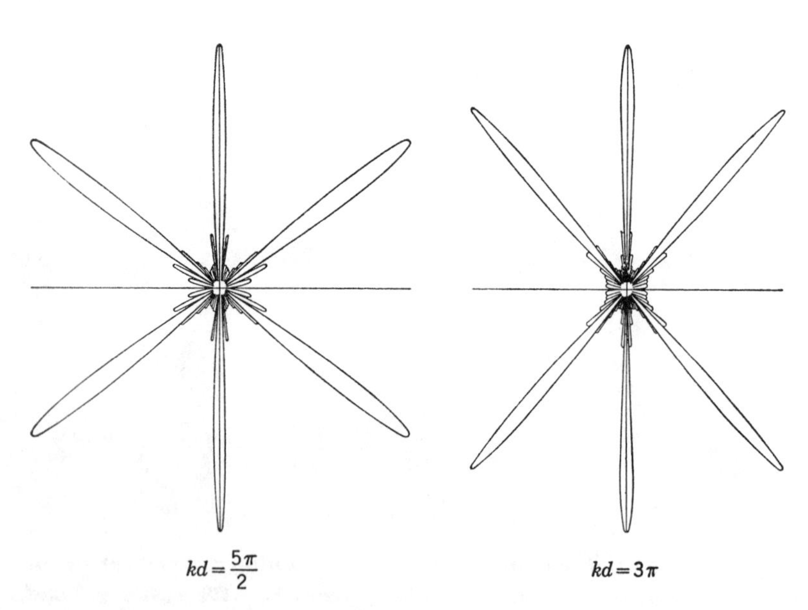

Fig. 3.8 Continued.

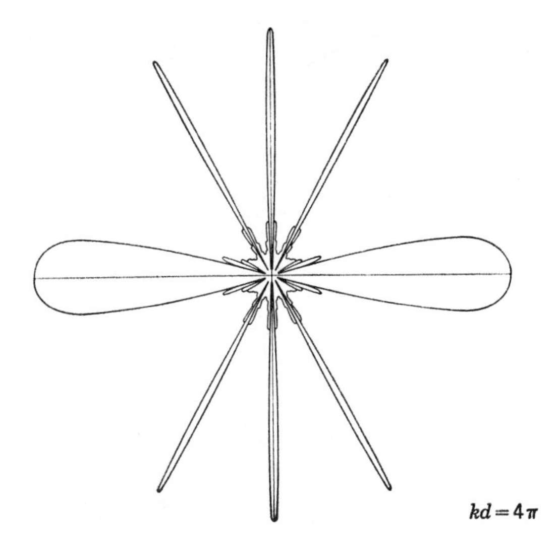

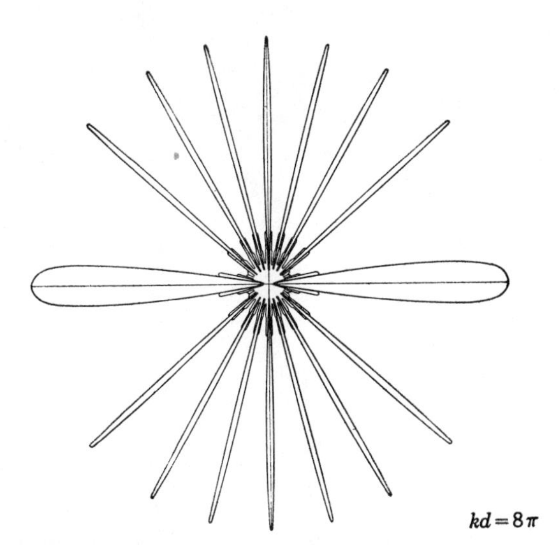

Fig. 3.8 Continued.

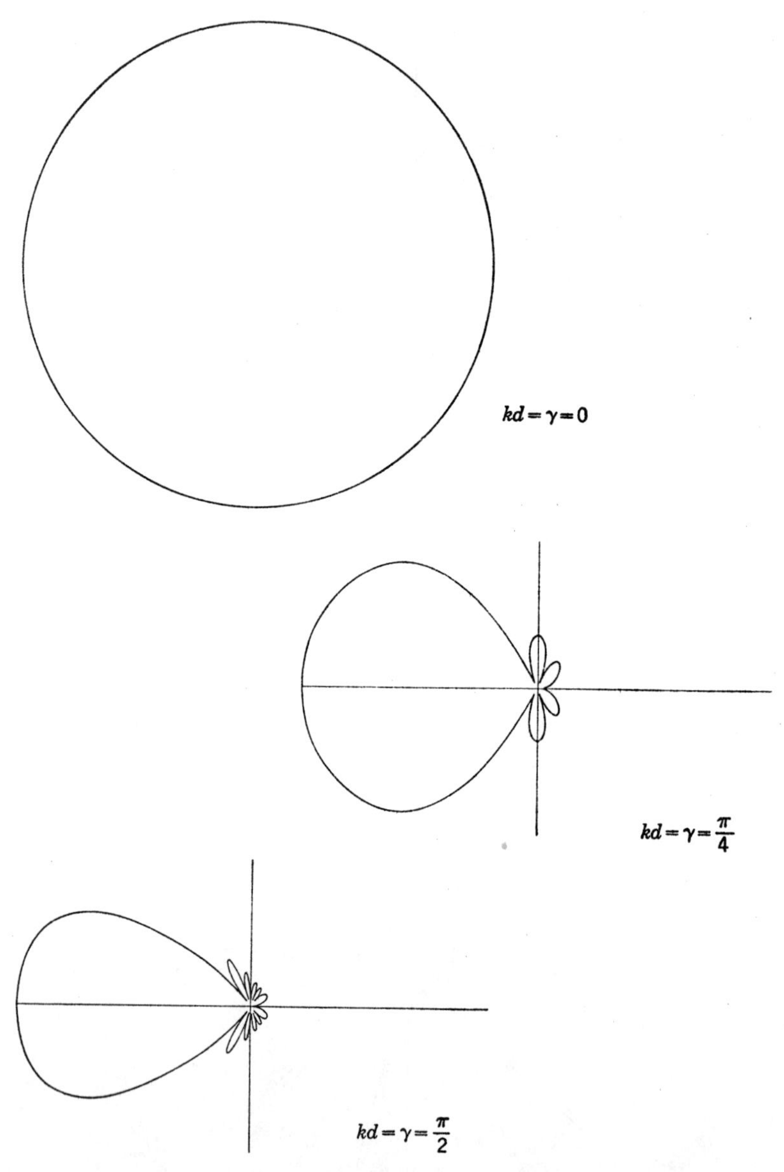

Fig. 3.9 *Radiation characteristic $K(\psi)$ of auniform linear array calculated from Eq. (80) with $n = 12$ for various values of $kd = \gamma$ Unilateral end-fire array. Bilateral end-fire array.*

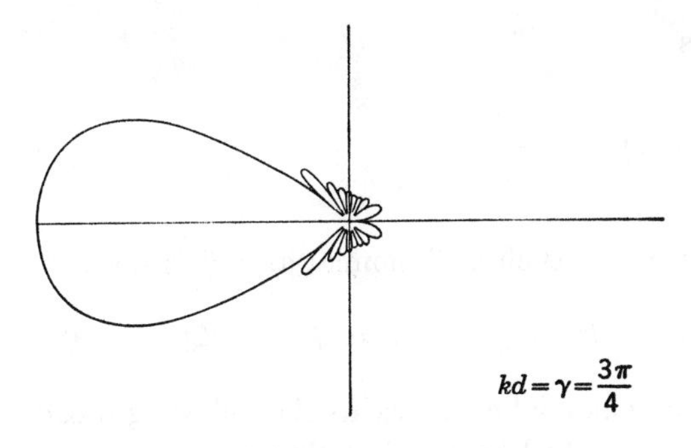

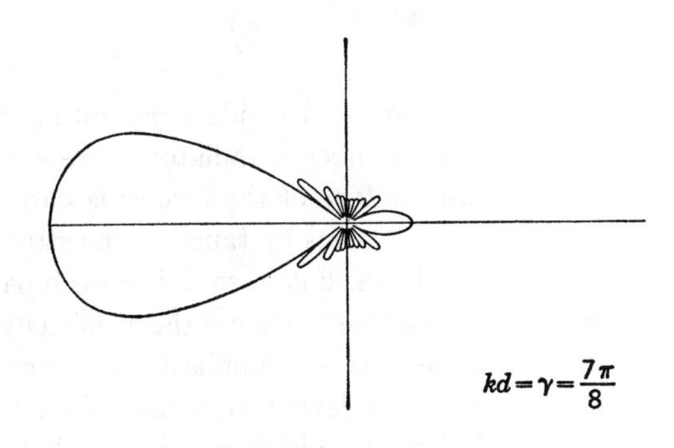

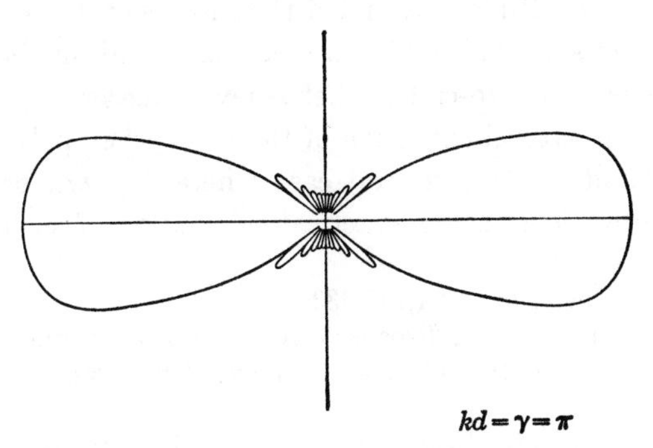

Fig. 3.9 Continued.

the array factor becomes

$$A(\psi) = \sum_{p=0}^{n-1} \binom{n-1}{p} \xi^p = (1+\xi)^{n-1} \tag{82}$$

Hence, the radiation pattern of such a "binomial array" is given by

$$|A(\psi)| = 2^{n-1}|\cos^{n-1}(\alpha/2)| = 2^{n-1}|\cos^{n-1}[(kd\cos\psi + \gamma)/2]| \tag{83}$$

For $\gamma = 0$ and $kd = \pi$, the binomial array yields the following broadside pattern, which is distinguished by the fact that it is free of side lobes:[1]

$$|A(\psi)| = 2^{n-1}\cos^{n-1}\left(\frac{\pi}{2}\cos\psi\right) \tag{84}$$

Comparing a uniform broadside array with a binomial broadside array having the same number of radiators, we see from the above examples that the broadside lobe of the former is narrower than the broadside lobe of the latter. Thus by tapering the strengths of the radiators we reduce the side lobes, but in so doing we broaden the broadside lobe. However, it is possible to choose the coefficients a_p such that the width of the broadside lobe is minimized for a fixed side-lobe level, or conversely, the side-lobe level is minimized for a fixed width of the broadside lobe. Indeed, Dolph[2] demonstrated that for the case in which the number of sources in the array is even and $d \geq \lambda/2$, such an optimum pattern can be achieved by matching the antenna polynomial (76) to a Chebyshev polynomial. Then Riblet[3] extended the discussion to the case in which the number of sources is odd and $d < \lambda/2$. And finally Pokrovskii,[4] through the use of the so-called Chebyshev-Akhiezer polynomials, which constitute a natural extension of the Chebyshev polynomials, succeeded in handling the general case where $d \geq \lambda/2$ or $d < \lambda/2$. Certain simplifications in the practical calculation of such

[1] J. S. Stone, U.S. Patents 1,643,323 and 1,715,433.

[2] C. L. Dolph, Current Distribution for Broadside Arrays which Optimize the Relationship between Beam Width and Side-lobe Level, *Proc. IRE*, **34:** 335 (1946).

[3] H. J. Riblet, Discussion on Dolph's Paper, *Proc. IRE*, **35:** 489 (1947).

[4] V. L. Pokrovskii, On Optimum Linear Antennas, *Radiotekhn. i Elektron.*, **1:** 593 (1956).

Chebyshev arrays were made by Barbiere[1] and by Van der Maas.[2] For a continuous distribution of isotropic radiators along a straight line, i.e., for a line source, the problem of an optimum broadside pattern (narrow beamwidth and low side lobes) was solved by T. T. Taylor.[3]

When the sources are incommensurably spaced, the point of departure is no longer the polynomial (76) but the more general expression (71). Clearly expression (71) is considerably more difficult to handle than expression (76), especially when the number of sources becomes large; but with the use of a computer, numerical results can be obtained in a straightforward manner. An unequally spaced array is generally more "broadband" than an equally spaced array, in the sense that its radiation pattern remains essentially unaltered over a broader band of operating frequencies. King, Packard, and Thomas[4] studied this attribute of unequally spaced arrays by numerically evaluating the radiation pattern for $A_p = 1$ and x_p chosen according to various spacing schemes. A general discussion of unequally spaced linear arrays has been reported by Unz,[5] and certain equivalences between equally and unequally spaced arrays have been noted by Sandler.[6]

Returning to the case of a uniform array whose radiation characteristic is given by expression (80), we see that if n and $kd(<\pi)$ are fixed and γ is varied from 0 to kd, the major lobe rotates from the broadside direction to the end-fire direction. This suggests that by continuously varying the phase γ the beam can be made to sweep continuously over an entire sector. It is on this principle that electrical scanning antennas operate.[7] The phases of the antennas are controlled elec-

[1] D. Barbiere, A Method for Calculating the Current Distribution of Tchebyscheff Arrays, *Proc. IRE*, **40:** 78 (1952).

[2] G. J. van der Maas, A Simplified Calculation for Dolph-Tchebyscheff Arrays, *J. Appl. Phys.*, **25:** 121 (1954).

[3] T. T. Taylor, Design of Line-source Antennas for Narrow Beamwidth and Low Side Lobes, *IRE Trans. Antennas Propagation*, **AP-3** (1): 16 (1955).

[4] D. D. King, R. F. Packard, and R. K. Thomas, Unequally-spaced Broadband Arrays, *IRE Trans. Antennas Propagation*, **AP-8** (4): 380 (1960).

[5] H. Unz, Linear Arrays with Arbitrarily Distributed Elements, Electronics Research Lab., series 60, issue 168, University of California, Berkeley, Nov. 2, 1956.

[6] S. S. Sandler, Some Equivalences between Equally and Unequally Spaced Arrays, *IRE Trans. Antennas Propagation*, **AP-8** (5): 496 (1960).

[7] For a review of the scanning properties of such arrays see, for example, W. H. von Aulock, Properties of Phased Arrays, *Proc. IRE*, **48:** 1715 (1960).

trically by phase shifters which form an integral part of the feed system. Although in many operational radars the scanning is done mechanically; electrical scanning is used in the case of large array antennas because it provides scanning patterns and scanning rates that cannot be obtained by mechanical means.

Without further calculation, we can deduce the radiation pattern of a rectangular array of dipoles. We do this by compounding the radiation pattern of a parallel array[1] with that of a collinear array. An expression for the radiation pattern of a collinear array of half-wave dipoles can be constructed from expressions (68) and (71). We note that the parallel array of Fig. 3.7 is transformed into a collinear array when the dipoles are rotated until their axes are aligned with the x axis. Clearly then, in view of expression (68), the radiation pattern of each rotated dipole is given by

$$F(\psi) = \frac{\cos\left(\frac{\pi}{2}\cos\psi\right)}{\sin\psi}$$

and the array factor remains the same as it was before the rotation, viz.,

$$A(\psi) = \sum_{p=0}^{n-1} A_p e^{-ikx_p \cos\psi}$$

Hence the radiation pattern of the collinear array turns out to be

$$U(\psi) = |F(\psi)A(\psi)| = \left| \frac{\cos\left(\frac{\pi}{2}\cos\psi\right)}{\sin\psi} \sum_{p=0}^{n-1} A_p e^{-ikx_p \cos\psi} \right| \tag{85}$$

It follows from this expression (by replacing ψ with θ) that the radiation pattern of a collinear array of dipoles lying along the z axis with dipole centers at the points z_p is given by

$$U(\theta) = \left| \frac{\cos\left(\frac{\pi}{2}\cos\theta\right)}{\sin\theta} \sum_{p=0}^{n-1} A_p e^{-ikz_p \cos\theta} \right| \tag{86}$$

[1] The linear array shown in Fig. 3.7 is called a "parallel array" whenever it becomes necessary to distinguish it from a collinear array.

By substituting this expression for $F(\theta)$ in Eq. (70) we get the radiation pattern of a rectangular array of half-wave dipoles which are parallel to the z axis and have centers at the points $x = x_p$, $z = z_q$ ($p = 0, 1, \ldots, n - 1$; $q = 0, 1, \ldots, m - 1$) in the xz plane. We can regard expression (86) as the radiation pattern of each element of the parallel array, i.e., we can replace $F(\theta)$ of expression (70) with $U(\theta)$ of expression (86), and thus obtain the following expression for the radiation pattern of a rectangular array of dipoles:[1]

$$U(\theta,\phi) = \frac{\cos\left(\frac{\pi}{2}\cos\theta\right)}{\sin\theta}\left|\sum_{p=0}^{n-1}\sum_{q=0}^{m-1} A_{pq}e^{-ikx_p \sin\theta\cos\phi}e^{-ikz_q\cos\theta}\right| \tag{87}$$

where A_{pq} denotes the complex magnitude of the current in the dipole at $x = x_p$, $z = z_q$. If the magnitudes of the dipole currents are equal to a constant, say I_0, and if the array constitutes a two-dimensional periodic lattice with uniform spacings d_x and d_z in the x and z directions ($x_p = pd_x$, $z_q = qd_z$), expression (87) reduces to

$$U(\theta,\phi) = I_0\frac{\cos\left(\frac{\pi}{2}\cos\theta\right)}{\sin\theta}\left|\frac{\sin\left[n(kd_x\sin\theta\cos\phi)/2\right]}{\sin\left[(kd_x\sin\theta\cos\phi)/2\right]}\frac{\sin\left[m(kd_z\cos\theta)/2\right]}{\sin\left[(kd_z\cos\theta)/2\right]}\right| \tag{88}$$

We see that such a rectangular array can cast a narrow beam in the direction ($\theta = \pi/2$, $\phi = \pi/2$) normal to the plane of the array. Along the axis of this beam at a distance $r = r_0$ from the array, the radial component of Poynting's vector is given by

$$S_r = \sqrt{\frac{\mu}{\epsilon}}\frac{I_0^2}{8\pi^2 r_0^2}U^2(\pi/2,\pi/2) = \sqrt{\frac{\mu}{\epsilon}}\frac{I_0^2 n^2 m^2}{8\pi^2 r_0^2} \tag{89}$$

or by

$$S_r = \sqrt{\frac{\mu}{\epsilon}}\frac{I_0^2}{8\pi^2 r_0^2}\frac{(L_x L_z)^2}{d_x^2 d_z^2} \tag{90}$$

[1] Although this expression was derived by considering a parallel array of similar collinear arrays, it is valid also for the more general case where the complex amplitudes A_{pq} of the dipole currents are arbitrarily chosen.

where $L_x(= nd_x)$ and $L_z(= md_z)$ are by definition the effective dimensions of the array.[1]

In view of expression (90), it appears that S_r increases quadratically with the area L_xL_z of the array. However, expression (89) is valid only for "small" or "moderately sized" arrays because as the array is enlarged the field at the fixed observation point ($r = r_0$, $\theta = \pi/2$, $\phi = \pi/2$) changes in nature from a far-zone, or Fraunhofer, field to a near-zone, or Fresnel, field. If we take $L(= \sqrt{L_x^2 + L_y^2})$ as the typical dimension of the array, the condition[2] that the array be contained well within the first Fresnel zone is

$$\sqrt{r_0^2 + L^2} - r_0 \leq \frac{\lambda}{4} \tag{91}$$

From this it follows that

$$L^2 \leq \frac{\lambda^2}{16} + \frac{\lambda r_0}{2} \tag{92}$$

Since $\lambda/r_0 \ll 1$, expression (92) reduces to

$$L^2 \leq \frac{\lambda}{2} r_0 \tag{93}$$

Thus we see that the critical value of L is $L_c = \sqrt{\lambda r_0/2}$. If $L < L_c$, the observation point is in the far zone and the previously derived formulas are valid. On the other hand, if $L > L_c$, the observation point is in the near zone and to find the radiation one must take into account the fact that the field is now of the Fresnel type. Tetelbaum[3]

[1] The effective dimensions so defined are the limiting values of the actual dimensions $(L_x)_{\text{actual}} = (n - 1)d_x$ and $(L_z)_{\text{actual}} = (m - 1)d_z + 2l$ as $n,m \to \infty$.

[2] Let $r_0^{-1} \cos \omega t$ be the field at observation point due to the dipole at origin. Then $(r_0^2 + L^2)^{-1/2} \cos (\omega t + \phi)$ is the field at observation point due to the farthest dipole. Assuming that $\sqrt{r_0^2 + L^2} \approx r_0$ in the denominator, we see that the resulting field is $\cos (\omega t) + \cos (\omega t + \phi) = A \cos (\omega t + \alpha)$. It follows that the modulus A is given by $A^2 = 2 + 2 \cos \phi$. The second term is positive as long as $\phi \leq \pi/2$, with $\phi \equiv (2\pi/\lambda)(\sqrt{r_0^2 + L^2} - r_0)$. Hence we have the condition $(2\pi/\lambda)(\sqrt{r_0^2 + L^2} - r_0) \leq \pi/2$, or $\sqrt{r_0^2 + L^2} - r_0 \leq \frac{\lambda}{4}$.

[3] S. Tetelbaum, On Some Problems of the Theory of Highly-directive Arrays, *J. Phys., Acad. Sci. U.S.S.R.*, **10:** 285 (1946).

has performed such a calculation for the case of a square array; his results show that as the array is made larger, S_r at first increases in accord with Eq. (90) and then behaves in a manner dictated by Cornu's spiral of Fresnel diffraction theory. A similar calculation has been made by Polk[1] for the case of a uniformly illuminated rectangular aperture antenna.

3.7 Directivity Gain

The directivity gain g of a directional antenna can be calculated from the relation

$$g = \frac{4\pi r^2 (S_r)_{\max}}{\int_0^{4\pi} S_r(r,\theta,\phi) r^2 \, d\Omega} \qquad (r \to \infty) \tag{94}$$

where $S_r(r,\theta,\phi)$ denotes the radial component of the far-zone Poynting vector, $(S_r)_{\max}$ the major-lobe maximum of $S_r(r,\theta,\phi)$, $d\Omega(= \sin\theta \, d\theta \, d\phi)$ the element of solid angle, and r the radius of a far-zone sphere. This relation directly yields $g = 1$ for an isotropic antenna and $g > 1$ for all other antennas.

Unless the antenna happens to be a short dipole or some other equally simple antenna, the problem of calculating directivity gain is complicated by the fact that the integral representing the time-average power P radiated by the antenna, viz.,

$$P = \int_0^{4\pi} S_r(r,\theta,\phi) r^2 \, d\Omega \qquad (r \to \infty) \tag{95}$$

cannot be evaluated by elementary means. The same difficulty arises in connection with the calculation of the radiation resistance R of an antenna,[2] because to find R from the definition $R = 2P/I^2$, where I is

[1] C. Polk, Optical Fresnel-zone Gain of a Rectangular Aperture, *IRE Trans. Antennas Propagation*, **AP-4** (1): 65–69 (1956).

[2] M. A. Bontsch-Bruewitsch, Die Strahlung der komplizierten rechtwinkeligen Antennen mit gleichbeschaffenen Vibratoren, *Ann. Phys.*, **81:** 425 (1926).

an arbitrary reference current, one is again faced with the task of calculating P. As an alternative, it is always possible to calculate P by Brillouin's emf method,[1] but the integral to which this method leads is generally as difficult to evaluate as the integral (95) posed by Poynting's vector method. The situation is eased considerably when the antenna is highly directional, for then $S_r(r,\theta,\phi)$ may be approximated by a function that simplifies the evaluation of the integral (95).

Let us first consider the simple case of a short wire antenna. From Eq. (19) we see that for $kl \ll 1$, the far-zone radial component of the Poynting vector has the form

$$S_r(r,\theta,\phi) = \frac{K}{r^2} \sin^2 \theta \tag{96}$$

where K is a constant that will drop out of the calculation due to the homogeneity of relation (94). The maximum of $S_r(r,\theta,\phi)$ occurs at $\theta = \pi/2$ and has the value

$$(S_r)_{\max} = \frac{K}{r^2} \tag{97}$$

Substituting expressions (96) and (97) into definition (94), we find that the gain of a short dipole is given by

$$g = \frac{2}{\int_0^\pi \sin^3 \theta \, d\theta} = \frac{3}{2} \tag{98}$$

As the antenna is lengthened, its gain increases moderately. To show this, we recall from Eq. (19) that for a center-driven antenna of arbitrary length the far-zone radial component of the Poynting vector has the form

$$S_r(r,\theta,\phi) = \frac{K}{r^2} \left[\frac{\cos (kl \cos \theta) - \cos kl}{\sin \theta} \right]^2 \tag{99}$$

where, now, $K = \sqrt{\mu/\epsilon}\, I_0{}^2/8\pi^2$. Substituting this expression into Eq. (95), we obtain the following integral representation for the time-

[1] A. A. Pistolkors, The Radiation Resistance of Beam Antennas, *Proc. IRE*, **17:** 562 (1929).

average power radiated by the antenna:

$$P = 2\pi K \int_0^\pi \frac{[\cos (kl \cos \theta) - \cos kl]^2}{\sin \theta} \, d\theta \tag{100}$$

To evaluate this integral we introduce the new variables $u(= kl \cos \theta)$ and $v(= kl - u)$. Thus

$$\int_0^\pi \frac{[\cos (kl \cos \theta) - \cos kl]^2}{\sin \theta} \, d\theta$$

$$= \frac{1}{2} \int_{-kl}^{kl} (\cos u - \cos kl)^2 \left(\frac{1}{kl - u} + \frac{1}{kl + u} \right) du$$

$$= \int_{-kl}^{kl} \frac{(\cos u - \cos kl)^2}{kl - u} \, du = \int_0^{2kl} \left[(1 + \cos 2kl)(1 - \cos v) \right.$$

$$\left. - \sin 2kl \left(\sin v - \frac{1}{2} \sin 2v \right) - \frac{\cos 2kl}{2} (1 - \cos 2v) \right] \frac{dv}{v}$$

and hence

$$P = 2\pi K \left[C + \ln 2kl - \text{Ci}\, 2kl + \frac{\sin 2kl}{2} (\text{Si}\, 4kl - 2\text{Si}\, 2kl) \right.$$

$$\left. + \frac{\cos 2kl}{2} (C + \ln kl + \text{Ci}\, 4kl - 2\text{Ci}\, 2kl) \right] \tag{101}$$

where

$$\text{Si}\, x = \int_0^x \frac{\sin \xi}{\xi} \, d\xi$$

is the sine integral

$$\text{Ci}\, x = -\int_x^\infty \frac{\cos \xi}{\xi} \, d\xi = C + \ln x - \int_0^x \frac{1 - \cos \xi}{\xi} \, d\xi$$

is the cosine integral, and $C(= 0.5722 \cdots)$ is Euler's constant. With the aid of a table of sine and cosine integrals,[1] P can be easily

[1] See, for example, E. Jahnke and F. Emde, "Tables of Functions," Dover Publications, Inc., New York, 1943.

computed from expression (101). In the case of a half-wave dipole ($kl = \pi/2$) it follows from Eqs. (99) and (101) that $(S_r)_{\max} = K/r^2$ and $P = 2\pi(1.22)K$. Inserting these results into Eq. (94), we find that the gain of a half-wave dipole[1] is $g = 1.64$. Similarly, in the case of a full-wave dipole we would find that $g = 2.53$. These examples illustrate that the gain of a linear antenna increases rather slowly with length, and to get really high gains from thin-wire antennas one must operate them in multielement arrays.

As an antenna of high-gain capabilities, let us now consider a uniform parallel array whose far-zone Poynting vector, in accord with Eqs. (67) and (79), has the radial component

$$S_r = \sqrt{\frac{\mu}{\epsilon}} \frac{1}{8\pi^2 r^2} \left| \frac{\cos\left(\frac{\pi}{2}\cos\theta\right)}{\sin\theta} \frac{\sin[n(kd\sin\theta\cos\phi + \gamma)/2]}{\sin[(kd\sin\theta\cos\phi + \gamma)/2]} \right|^2 \tag{102}$$

By virtue of approximation (25) we can write this expression in the simpler form

$$S_r = \sqrt{\frac{\mu}{\epsilon}} \frac{(0.945)^2}{8\pi^2 r^2} \left| \sin\theta \frac{\sin[n(kd\sin\theta\cos\phi + \gamma)/2]}{\sin[(kd\sin\theta\cos\phi + \gamma)/2]} \right|^2 \tag{103}$$

It is clear that the maximum of S_r occurs at $\theta = \pi/2$ and $\phi = \phi_0$, where ϕ_0 is fixed by $kd\cos\phi_0 + \gamma = 0$. Thus

$$(S_r)_{\max} = \sqrt{\frac{\mu}{\epsilon}} \frac{(0.945)^2}{8\pi^2 r^2} n^2 \tag{104}$$

Substituting (103) and (104) into definition (94), we get

$$g = \frac{4\pi n^2}{\displaystyle\int_0^{2\pi}\int_0^{\pi} \sin^3\theta \frac{\sin^2[n(kd\sin\theta\cos\phi + \gamma)/2]}{\sin^2[(kd\sin\theta\cos\phi + \gamma)/2]} \, d\theta \, d\phi} \tag{105}$$

The integral in this expression can be evaluated exactly[2] through the

[1] The relative gain g_r of an antenna is its gain over a half-wave dipole. That is, $g_r = g/1.64$.

[2] C. H. Papas and R. King, The Radiation Resistance of End-fire and Collinear Arrays, *Proc. IRE*, **36:** 736 (1948).

use of Sonine's first integral theorem[1]

$$\int_0^{2\pi}\int_0^{\pi} \sin^3\theta \frac{\sin^2[n(kd\sin\theta\cos\phi+\gamma)/2]}{\sin^2[(kd\sin\theta\cos\phi+\gamma)/2]}\,d\theta d\phi$$

$$= \frac{8\pi n}{3} + 8\pi \sum_{q=1}^{n-1} (n-q)\cos(q\gamma)\left(\frac{\sin u}{u} - \frac{\sin u}{u^3} + \frac{\cos u}{u^2}\right) \qquad (106)$$

where $u = qkd$. Hence the gain (105) of the array can be expressed in terms of the finite series (106):

$$g = \frac{4\pi n^2}{\dfrac{8\pi n}{3} + 8\pi \displaystyle\sum_{q=1}^{n-1} (n-q)\cos(q\gamma)\left(\frac{\sin u}{u} - \frac{\sin u}{u^3} + \frac{\cos u}{u^2}\right)} \qquad (107)$$

This expression is convenient for numerical calculation, especially when the number n of dipoles is small. When n is very large and the array is operating as a broadside array ($\gamma = 0$, $d \leq \lambda$), we have the simple limiting form

$$g = \frac{4nd}{\lambda} \qquad (n \to \infty) \qquad (108)$$

which may be obtained[2] by comparing the denominator of expression (107) with the Fourier expansions of the functions x, x^2, and x^3 for the interval $(0,2\pi)$.

Let us now calculate the gain of a large uniform rectangular array. We recall from Eq. (88) that its radiation pattern is given by

$$U(\theta,\phi) = I_0 \frac{\cos\left(\dfrac{\pi}{2}\cos\theta\right)}{\sin\theta}\left|\frac{\sin n\alpha}{\sin\alpha}\frac{\sin m\beta}{\sin\beta}\right| \qquad (109)$$

where $\alpha = (kd_x/2)\sin\theta\cos\phi$ and $\beta = (kd_z/2)\cos\theta$. The spacings d_x and d_z are assumed to be less than a wavelength ($d_x < \lambda$, $d_z < \lambda$) and

[1] N. J. Sonine, Recherches sur les fonctions cylindriques et le développement des fonctions continues en séries, *Math. Ann.*, **16:** 1 (1880).

[2] See, for example, K. Franz and H. Lassen, "Antennen und Ausbreitung," p. 255, Springer-Verlag OHG, Berlin, 1956.

hence the radiation pattern consists of two broadside beams, one in the direction $\theta = \pi/2$, $\phi = \pi/2$ and the other on the opposite side of the array in the direction $\theta = \pi/2$, $\phi = 3\pi/2$. Each of these beams has the maximum value

$$U_{\max} = I_0 nm \tag{110}$$

With the aid of expressions (109) and (110) and the fact that S_r is proportional to U^2/r^2, definition (94) leads to the following expression for the gain:

$$g = \frac{4\pi n^2 m^2}{\displaystyle\int_0^{2\pi}\int_0^{\pi} \frac{\cos^2\left(\frac{\pi}{2}\cos\theta\right)}{\sin\theta}\,\frac{\sin^2 n\alpha}{\sin^2\alpha}\,\frac{\sin^2 m\beta}{\sin^2\beta}\,d\theta d\phi} \tag{111}$$

Since the array is large, most of its radiation is concentrated in the two narrow broadside beams. Because of the symmetry of the radiation pattern, the ϕ integration may be restricted to the beam lying in the interval $(0,\pi)$, and because of the sharpness of the beam, the following approximations obtain:

$$\sin\theta \approx 1 \qquad \cos\phi \approx \pi/2 - \phi \qquad \sin\phi \approx 1$$

$$d\alpha = (kd_x/2)(\cos\theta\cos\phi\,d\theta - \sin\theta\sin\phi\,d\phi) \approx -(kd_x/2)d\phi$$

$$d\beta = -(kd_z/2)\sin\theta\,d\theta \approx -(kd_z/2)d\theta$$

Applying these approximations to the integral in Eq. (111), we get

$$g = \frac{2\pi n^2 m^2 (kd_x/2)(kd_z/2)}{\displaystyle\int_{-\infty}^{\infty}\int_{-\infty}^{\infty} \frac{\sin^2 n\alpha}{\alpha^2}\,\frac{\sin^2 m\beta}{\beta^2}\,d\alpha d\beta} \tag{112}$$

Here the actual limits have been replaced by infinite ones on the ground that the two factors in the integrand rapidly decrease as α and β depart from zero. Since the α integration yields $n\pi$ and the β integration $m\pi$, expression (112) yields the following limiting value for the gain of the rectangular array:[1]

$$g = \frac{2\pi(nd_x)(md_z)}{\lambda^2} \qquad (n, m \to \infty) \tag{113}$$

If the array were backed by a reflector, which eliminates one of the beams and concentrates all the energy in the other, the limiting value of the gain would be twice as large, viz.,

$$g = \frac{4\pi(nd_x)(md_z)}{\lambda^2} \qquad (n, m \to \infty) \tag{114}$$

In terms of the effective dimensions of the array $L_x(= nd_x)$ and $L_z(= md_z)$ and the effective area of the array $A(= L_xL_z)$, the limiting values (113) and (114) respectively become

$$g = 2\pi \frac{L_xL_z}{\lambda^2} = 2\pi \frac{A}{\lambda^2} \tag{115}$$

[1] If the limits are chosen to include only the broadside beam, then the integral in Eq. (112) must be replaced by

Integral =

$$\int_{-\pi/m}^{\pi/m} \int_{-\pi/n}^{\pi/n} \frac{\sin^2 nd}{\alpha^2} \frac{\sin^2 m\beta}{\beta^2} \, d\alpha d\beta = nm \int_{-\pi}^{\pi} \int_{-\pi}^{\pi} \frac{\sin^2 x}{x^2} \frac{\sin^2 y}{y^2} \, dx dy$$

Since

$$\frac{d}{d\xi}\left(\frac{\sin^2 \xi}{\xi}\right) = \frac{-\sin^2 \xi}{\xi^2} + \frac{\sin 2\xi}{\xi}$$

then

$$\int_{-\pi}^{\pi} \frac{\sin^2 \xi}{\xi^2} \, d\xi = \frac{-\sin^2 \xi}{\xi}\Bigg|_{-\pi}^{\pi} + \int_{-\pi}^{\pi} \frac{\sin 2\xi}{\xi} \, d\xi$$

The first term on the right vanishes and the second term is equal to 2Si (2π). Using this result, we get Integral $= 4nm[\mathrm{Si}\ (2\pi)]^2$. Hence the corresponding expression for the gain is $g = \dfrac{2\pi(nd_x)(md_z)}{\gamma^2}\left[\dfrac{\pi}{2\mathrm{Si}\ (2\pi)}\right]^2$. This result agrees with Eq. (113) since $[\pi/2\mathrm{Si}\ (2\pi)]^2$ is approximately equal to 1.

for the array without a reflector, and

$$g = 4\pi \frac{L_x L_z}{\lambda^2} = 4\pi \frac{A}{\lambda^2} \tag{116}$$

for the array with a reflector.

It is clear from the above results that the gain of an array can be increased by increasing its size. However, it is also possible in principle to achieve very high gain, i.e., supergain, with an array of limited dimensions.[1] Since the elements of such superdirective arrays are closely spaced, their mutual interactions play a determining role. These interactions have the effect of storing reactive energy in the neighborhood of the elements and of thus making narrow the bandwidth of the array.[2] Moreover, the large currents that superdirectivity demands lead to high ohmic losses and consequently to reductions in operating efficiency.[3] In addition to narrow bandwidths and low efficiencies, superdirective antennas are burdened with the requirement that the amplitudes and phases of the currents be maintained with a relatively high degree of precision. Superdirective arrays are useful in those cases where a very sharp narrow beam is desired regardless of the cost in bandwidth, efficiency, and critical tolerances. Some aspects of the supergain phenomenon are closely related to the problem of optical resolving power.[4]

[1] S. A. Schelkunoff, A Mathematical Theory of Linear Arrays, *Bell System Tech. J.*, **22:** 80 (1943).

[2] L. J. Chu, Physical Limitations of Omni-directional Antennas, *J. Appl. Phys.*, **19:** 1163 (1948).

[3] R. M. Wilmotte, Note on Practical Limitations in the Directivity of Antennas, *Proc. IRE*, **36:** 878 (1948); T. T. Taylor, A Discussion on the Maximum Directivity of an Antenna, *Proc. IRE*, **36:** 1135 (1948); H. J. Riblet, Note on Maximum Directivity of an Antenna, *Proc. IRE*, **36:** 620 (1948).

[4] G. Toraldo di Francia, Directivity, Super-gain and Information Theory, *IRE Trans. Antennas Propagation*, **AP-4** (3): 473 (1956).

Multipole expansion of the radiation field 4

One method of expanding a radiation field in multipoles is to develop in Taylor series the Helmholtz integral representations of the scalar and vector potentials and then to identify the terms of the series with formal generalizations of the conventional multipoles of electrostatics and magnetostatics. Another method of expansion consists in developing the radiation field in spherical E and H waves and defining the E waves as electric multipole fields and the H waves as magnetic multipole fields. In this chapter a brief account is given of these two methods.

4.1 Dipole and Quadrupole Moments

We assume that a monochromatic current density $\mathbf{J}(\mathbf{r}')$ is distributed throughout some bounded region of space. Then by virtue of the conservation of charge there also exists in the region a monochromatic charge density $\rho(\mathbf{r}')$ given by

$$\nabla' \cdot \mathbf{J}(\mathbf{r}') = i\omega\rho(\mathbf{r}') \tag{1}$$

To deduce a relation which we will use for defining the moments of the charge in terms of the current and for framing the gauge of the potentials produced by the charge and current, we multiply this equation of continuity by an arbitrary function $f(\mathbf{r},\mathbf{r}')$ and then integrate with respect to the primed coordinates. Thus from Eq. (1) we obtain

$$\int \rho(\mathbf{r}')f(\mathbf{r},\mathbf{r}')dV' = \frac{1}{i\omega}\int f(\mathbf{r},\mathbf{r}')\nabla'\cdot\mathbf{J}(\mathbf{r}')dV' \qquad (2)$$

Here the region of integration includes the entire space occupied by the current and charge, and the normal component of the current is zero on the surface which bounds the region. Using the identity $\nabla'\cdot(\mathbf{J}f) = f\nabla'\cdot\mathbf{J} + \mathbf{J}\cdot\nabla'f$, and noting that the term $\int\nabla'\cdot(\mathbf{J}f)dV'$ disappears because by the divergence theorem it equals the surface integral $\int f\mathbf{n}\cdot\mathbf{J}\,dS'$ whose integrand disappears, we see that Eq. (2) leads to the desired relation

$$\int \rho(\mathbf{r}')f(\mathbf{r},\mathbf{r}')dV' = \frac{i}{\omega}\int \mathbf{J}(\mathbf{r}')\cdot\nabla'f(\mathbf{r},\mathbf{r}')dV' \qquad (3)$$

On the proper selection of $f(\mathbf{r},\mathbf{r}')$, the left side of this relation becomes a moment of the charge and the right side becomes an equivalent representation of the moment in terms of the current.

When $f = 1$, relation (3) reduces to

$$\int\rho(\mathbf{r}')dV' = 0 \qquad (4)$$

and we thus see that the total charge is zero. Moreover, when we denote the cartesian components of $\mathbf{r}'$ and $\mathbf{J}$ by x'_α and J_α ($\alpha = 1, 2, 3$), and when we successively assume that $f = x'_\alpha$ and $f = x'_\alpha x'_\beta$ ($\alpha, \beta = 1, 2, 3$), relation (3) gives rise to the first and second moments of the charge, which define respectively the cartesian components p_α and $Q_{\alpha\beta}$ of the electric dipole moment $\mathbf{p}$ and the electric quadrupole moment $\mathbf{Q}$:

$$p_\alpha = \int \rho(\mathbf{r}')x'_\alpha\,dV' = \frac{i}{\omega}\int J_\alpha(\mathbf{r}')dV' \qquad (\alpha = 1, 2, 3) \qquad (5)$$

$$\begin{aligned} Q_{\alpha\beta} &= \int \rho(\mathbf{r}')x'_\alpha x'_\beta\,dV' \\ &= \frac{i}{\omega}\int \left[J_\alpha(\mathbf{r}')x'_\beta + J_\beta(\mathbf{r}')x'_\alpha\,dV'\right] \qquad (\alpha, \beta = 1, 2, 3) \end{aligned} \qquad (6)$$

It is clear from these expressions that **p** is a vector and **Q** is a dyadic whose components constitute a symmetrical matrix, $Q_{\alpha\beta} = Q_{\beta\alpha}$. In vector form, Eqs. (5) and (6) are

$$\mathbf{p} = \int \rho(\mathbf{r}')\mathbf{r}'\, dV' = \frac{i}{\omega}\int \mathbf{J}(\mathbf{r}')dV' \tag{7}$$

$$\mathbf{Q} = \int \rho(\mathbf{r}')\mathbf{r}'\mathbf{r}'\, dV' = \frac{i}{\omega}\int [\mathbf{J}(\mathbf{r}')\mathbf{r}' + \mathbf{r}'\mathbf{J}(\mathbf{r}')]dV' \tag{8}$$

These relations show how **p** and **Q** can be calculated from a knowledge of either the charge or the current.

Associated with the electric charge and electric current are their magnetic counterparts, the magnetic charge density $\rho_m(\mathbf{r}')$ and magnetic current density $\mathbf{J}_m(\mathbf{r}')$. These conceptual entities serve the purpose of establishing a formal duality between electric and magnetic quantities. The magnetic current density is defined by $\mathbf{J}_m(\mathbf{r}') = (\omega/2i)\mathbf{r}' \times \mathbf{J}(\mathbf{r}')$ and the magnetic charge density is deduced in turn from $\mathbf{J}_m(\mathbf{r}')$ by the conservation law $\nabla' \cdot \mathbf{J}_m(\mathbf{r}') = i\omega\rho_m(\mathbf{r}')$. Since $\mathbf{J}_m(\mathbf{r}')$ and $\rho_m(\mathbf{r}')$ obey the conservation law, it follows that they also obey a relation which is formally the same as Eq. (3), viz.,

$$\int \rho_m(\mathbf{r}')f(\mathbf{r},\mathbf{r}')dV' = \frac{i}{\omega}\int \mathbf{J}_m(\mathbf{r}') \cdot \nabla' f(\mathbf{r},\mathbf{r}')dV' \tag{9}$$

Choosing f to be successively the cartesian components of $\mathbf{r}'$, and recalling the definition of $\mathbf{J}_m(\mathbf{r}')$, we get the following expression for the first moment of the magnetic charge density:

$$\int \rho_m(\mathbf{r}')\mathbf{r}'\, dV' = \frac{i}{\omega}\int \mathbf{J}_m(\mathbf{r}')dV' = \frac{1}{2}\int \mathbf{r}' \times \mathbf{J}(\mathbf{r}')dV' \tag{10}$$

Since the first moment of $\rho_m(\mathbf{r}')$ is by definition the magnetic dipole moment **m**, this equation gives the following expression for **m** in terms of the electric current:

$$\mathbf{m} = \tfrac{1}{2}\int \mathbf{r}' \times \mathbf{J}(\mathbf{r}')dV' \tag{11}$$

Clearly **m** can be regarded as a pseudo vector whose cartesian components are[1]

$$m_\gamma = \frac{1}{2} \sum_{\alpha,\beta=1}^{3} \epsilon_{\alpha\beta\gamma} \int x'_\alpha J_\beta(r') dV' \qquad (\gamma = 1, 2, 3) \tag{12}$$

or as an antisymmetrical dyadic $\mathbf{m}$ having the cartesian components

$$m_{\beta\alpha} = \tfrac{1}{2} \int [x'_\alpha J_\beta(\mathbf{r}') - x'_\beta J_\alpha(\mathbf{r}')] dV' \qquad (\alpha, \beta = 1, 2, 3) \tag{13}$$

The simplest current configuration that possesses an electric dipole moment is the short filament of current

$$\mathbf{J} = \mathbf{e}_z I_0 \delta(x')\delta(y') \qquad (-l \leq z' \leq l) \tag{14}$$

Substituting this expression into Eqs. (7), (8), (11) we find that the electric dipole moment is given by

$$\mathbf{p} = \mathbf{e}_z \frac{i}{\omega} 2lI_0 \tag{15}$$

and the electric quadrupole and magnetic dipole moments are zero. The dual of this configuration is the small filamentary loop of current

$$\mathbf{J} = \mathbf{e}_\phi J_\phi = \mathbf{e}_\phi I_0 \delta(\rho' - a)\delta(z') \tag{16}$$

where a is the radius of the loop. Substituting into Eqs. (7), (8), (11) we find in this case that the only nonzero moment is the magnetic dipole moment given by

$$\mathbf{m} = \frac{I_0}{2} \mathbf{e}_\rho \times \mathbf{e}_\phi \int \rho' \delta(\rho' - a)\delta(z')\rho' d\rho' d\phi' dz' = \mathbf{e}_z \pi a^2 I_0 \tag{17}$$

As an example of a configuration having an electric quadrupole moment,

[1] The three-index symbol $\epsilon_{\alpha\beta\gamma}$ has the following meaning: $\epsilon_{\alpha\beta\gamma} = 0$ when any two of the subscripts are the same, $\epsilon_{\alpha\beta\gamma} = 1$ when α, β, γ are all different and occur in the order 12312 · · · (even permutations of 123), and $\epsilon_{\alpha\beta\gamma} = -1$ when α, β, γ are all different and occur in the order 21321 · · · (odd permutations of 123). That is, $\epsilon_{123} = \epsilon_{231} = \epsilon_{312} = 1$ and $\epsilon_{213} = \epsilon_{132} = \epsilon_{321} = -1$.

let us take two antiparallel short filaments of current separated by a distance d:

$$\mathbf{J} = \mathbf{e}_z I_0[\delta(x' - d/2)\delta(y') - \delta(x' + d/2)\delta(y')] \qquad (-l \leq z' \leq l) \tag{18}$$

Substituting in Eq. (8), we get

$$\mathbf{Q} = \frac{i}{\omega} 2lI_0(\mathbf{e}_z\mathbf{e}_x + \mathbf{e}_x\mathbf{e}_z) \int [x'\delta(x' - d/2) - x'\delta(x' + d/2)]dx'$$

or

$$\mathbf{Q} = \left(\frac{i}{\omega} 2lI_0\right)(\mathbf{e}_z\mathbf{e}_x + \mathbf{e}_x\mathbf{e}_z)d \tag{19}$$

Since the dipole moment of each filament is given by Eq. (15) and $\mathbf{d} = d\mathbf{e}_x$ is the vector separation of the two filaments, we can write the electric quadrupole moment as

$$\mathbf{Q} = \mathbf{pd} + \mathbf{dp} \tag{20}$$

If we choose the function $f(\mathbf{r},\mathbf{r}')$ in relation (3) to be the free-space Green's function, i.e., if we let

$$f(\mathbf{r},\mathbf{r}') = \frac{e^{ik|\mathbf{r}-\mathbf{r}'|}}{4\pi|\mathbf{r} - \mathbf{r}'|} \tag{21}$$

then by virtue of the identity

$$\nabla' \frac{e^{ik|\mathbf{r}-\mathbf{r}'|}}{|\mathbf{r} - \mathbf{r}'|} = -\nabla \frac{e^{ik|\mathbf{r}-\mathbf{r}'|}}{|\mathbf{r} - \mathbf{r}'|} \tag{22}$$

relation (3) yields

$$\int \rho(\mathbf{r}') \frac{e^{ik|\mathbf{r}-\mathbf{r}'|}}{4\pi|\mathbf{r} - \mathbf{r}'|} dV' = -\frac{i}{\omega} \nabla \cdot \int \mathbf{J}(\mathbf{r}') \frac{e^{ik|\mathbf{r}-\mathbf{r}'|}}{4\pi|\mathbf{r} - \mathbf{r}'|} dV' \tag{23}$$

The left side of this equation is the Helmholtz integral representation of $\epsilon\phi(\mathbf{r})$; the integral on the right is the Helmholtz integral representation of $(1/\mu)\mathbf{A}(\mathbf{r})$. Hence, Eq. (23) expresses the Lorentz condition coupling the vector potential $\mathbf{A}(\mathbf{r})$ with the scalar potential $\phi(\mathbf{r})$, viz.,

$$\nabla \cdot \mathbf{A}(\mathbf{r}) = i\omega\epsilon\mu\phi(\mathbf{r}) \tag{24}$$

Thus we see that the Lorentz gauge is the one that is consistent with the conservation of charge.

4.2 Taylor Expansion of Potentials

As in the previous section, let us start by assuming that in a bounded region of space we have an arbitrary distribution of monochromatic current density $\mathbf{J}(\mathbf{r}')$ and a distribution of charge density $\rho(\mathbf{r}')$ derived from it by the equation of continuity. Then the scalar and vector potentials of the electromagnetic field produced by such a source are given by the Helmholtz integrals

$$\phi(\mathbf{r}) = \frac{1}{4\pi\epsilon}\int \rho(\mathbf{r}')\,\frac{e^{ik|\mathbf{r}-\mathbf{r}'|}}{|\mathbf{r}-\mathbf{r}'|}\,dV' \tag{25}$$

$$\mathbf{A}(\mathbf{r}) = \frac{\mu}{4\pi}\int \mathbf{J}(\mathbf{r}')\,\frac{e^{ik|\mathbf{r}-\mathbf{r}'|}}{|\mathbf{r}-\mathbf{r}'|}\,dV' \tag{26}$$

To expand these potentials in Taylor series we need the three-dimensional generalization of the familiar one-dimensional Taylor series. We recall that the one-dimensional Taylor series expansion of a function $f(x)$ is given by

$$f(x+h) = \sum_{n=0}^{\infty}\frac{1}{n!}\left(h\frac{d}{dx}\right)^n f(x) \tag{27}$$

When x is replaced by $\mathbf{r}$, h by some vector $\mathbf{a}$, and hd/dx by $\mathbf{a}\cdot\nabla$, the Taylor series (27) heuristically takes the three-dimensional form

$$f(\mathbf{r}+\mathbf{a}) = \sum_{n=0}^{\infty}\frac{1}{n!}(\mathbf{a}\cdot\nabla)^n f(\mathbf{r}) \tag{28}$$

By letting $\mathbf{a} = -\mathbf{r}'$ we see that for any function of $\mathbf{r} - \mathbf{r}'$ the expansion is

$$f(\mathbf{r} - \mathbf{r}') = \sum_{n=0}^{\infty} \frac{1}{n!} (-\mathbf{r}' \cdot \nabla)^n f(\mathbf{r}) \tag{29}$$

and hence we have

$$\frac{e^{ik|\mathbf{r}-\mathbf{r}'|}}{|\mathbf{r} - \mathbf{r}'|} = \sum_{n=0}^{\infty} \frac{1}{n!} (-\mathbf{r}' \cdot \nabla)^n \frac{e^{ikr}}{r} \tag{30}$$

With the aid of expansion (30) we develop the Helmholtz integral representation (25) of the scalar potential in the Taylor series

$$\phi(\mathbf{r}) = \frac{1}{4\pi\epsilon} \int \rho(\mathbf{r}')[1 - \mathbf{r}' \cdot \nabla + \tfrac{1}{2}(\mathbf{r}' \cdot \nabla)^2 - \cdots] \frac{e^{ikr}}{r} \, dV' \tag{31}$$

Keeping only the first three terms, we get

$$4\pi\epsilon\phi(\mathbf{r}) = \left[\int \rho(\mathbf{r}')dV'\right] \frac{e^{ikr}}{r} - \left[\int \rho(\mathbf{r}')\mathbf{r}' \, dV'\right] \cdot \nabla \frac{e^{ikr}}{r} + \tfrac{1}{2}\left[\int \rho(\mathbf{r}')\mathbf{r}'\mathbf{r}' \, dV'\right] : \nabla\nabla \frac{e^{ikr}}{r} \tag{32}$$

where in the third term $(\mathbf{r}' \cdot \nabla)^2$ has been written as the double scalar product $\mathbf{r}'\mathbf{r}' : \nabla\nabla$ of the dyadics $\mathbf{r}'\mathbf{r}'$ and $\nabla\nabla$. The first term is zero by virtue of Eq. (4). The second and third terms involve respectively the electric dipole and quadrupole moments, as defined by Eqs. (7) and (8). Thus the leading terms of the Taylor expansion of $\phi(\mathbf{r})$ can be written in the following concise form:

$$\phi(\mathbf{r}) = -\frac{1}{4\pi\epsilon} \left(\mathbf{p} \cdot \nabla \frac{e^{ikr}}{r} - \tfrac{1}{2}\mathbf{Q} : \nabla\nabla \frac{e^{ikr}}{r} + \cdots \right) \tag{33}$$

where $\mathbf{p}$ is the electric dipole moment and $\mathbf{Q}$ the electric quadrupole moment.

Similarly, the Taylor series development of the Helmholtz integral representation (26) for the vector potential is

$$4\pi\mathbf{A}(\mathbf{r}) = \mu\int\mathbf{J}(\mathbf{r}')[1 - \mathbf{r}'\cdot\nabla + \tfrac{1}{2}(\mathbf{r}'\cdot\nabla)^2 - \cdots]\frac{e^{ikr}}{r}\,dV' \tag{34}$$

Considering only the first two terms, we get

$$(4\pi/\mu)\mathbf{A}(\mathbf{r}) = \left[\int\mathbf{J}(\mathbf{r}')dV'\right]\frac{e^{ikr}}{r} - \left[\int\mathbf{J}(\mathbf{r}')\mathbf{r}'\,dV'\right]\cdot\nabla\frac{e^{ikr}}{r} \tag{35}$$

The first integral by Eq. (7) is equal to $-i\omega\mathbf{p}$. To express the second integral in terms of $\mathbf{Q}$ and $\mathbf{m}$, we decompose the dyadic $\mathbf{J}(\mathbf{r}')\mathbf{r}'$ into its symmetric and antisymmetric parts:

$$\mathbf{J}(\mathbf{r}')\mathbf{r}' = \tfrac{1}{2}[\mathbf{r}'\mathbf{J}(\mathbf{r}') + \mathbf{J}(\mathbf{r}')\mathbf{r}'] - \tfrac{1}{2}[\mathbf{r}'\mathbf{J}(\mathbf{r}') - \mathbf{J}(\mathbf{r}')\mathbf{r}']$$

When integrated, the first symmetric part yields by Eq. (8) the symmetric dyadic $(-i\omega/2)\mathbf{Q}$ and the second antisymmetric part yields by Eq. (13) the antisymmetric dyadic $\mathbf{m}$. Thus the expansion for the vector potential up to the second term turns out to be

$$4\pi\mathbf{A}(\mathbf{r}) = -i\omega\mu\mathbf{p}\frac{e^{ikr}}{r} + \frac{i\omega\mu}{2}\mathbf{Q}\cdot\nabla\frac{e^{ikr}}{r} - \mu\mathbf{m}\cdot\nabla\frac{e^{ikr}}{r} \tag{36}$$

If one prefers to think of $\mathbf{m}$ as a pseudo vector, then the operator $\mathbf{m}\cdot\nabla$, where $\mathbf{m}$ is an antisymmetric dyadic, has to be replaced by $\mathbf{m}\times\nabla$, where $\mathbf{m}$ is a pseudo vector. Accordingly, an alternative form of expression (36) is

$$4\pi\mathbf{A}(\mathbf{r}) = -i\omega\mu\mathbf{p}\frac{e^{ikr}}{r} + \frac{i\omega\mu}{2}\mathbf{Q}\cdot\nabla\frac{e^{ikr}}{r} - \mu\mathbf{m}\times\nabla\frac{e^{ikr}}{r} \tag{37}$$

Hence, for a source that can be described as a superposition of an electric dipole, a magnetic dipole, and an electric quadrupole, the scalar and vector potentials are given by Eqs. (33) and (37). If a source is such that poles of higher multiplicity are required, it becomes more convenient to calculate $\phi(\mathbf{r})$ and $\mathbf{A}(\mathbf{r})$ by evaluating directly the Helmholtz integrals than to use the method of multipole expansion.

4.3 Dipole and Quadrupole Radiation

The electromagnetic field of a monochromatic source can be found by substituting the Taylor expansions of the scalar and vector potentials, viz.,

$$\phi(\mathbf{r}) = -\frac{1}{\epsilon}(\mathbf{p}\cdot\nabla G - \tfrac{1}{2}\mathbf{Q}:\nabla\nabla G + \cdots) \tag{38}$$

$$\mathbf{A}(\mathbf{r}) = -i\omega\mu(\mathbf{p}G - \tfrac{1}{2}\mathbf{Q}\cdot\nabla G - \frac{i}{\omega}\mathbf{m}\times\nabla G + \cdots) \tag{39}$$

where $G = e^{ikr}/4\pi r$, into the relations

$$\mathbf{E} = -\nabla\phi + i\omega\mathbf{A} \qquad \mathbf{H} = (1/\mu)\nabla\times\mathbf{A} \tag{40}$$

which yield the electromagnetic field **E, H.** By virtue of the linearity of the system, the resulting electromagnetic field may be thought of as the vector sum of the individual electromagnetic fields of the various poles. Since each multipole radiates a spherical wave and the most natural coordinate system for a mathematical description of the radiation is a spherical one centered on the multipoles, we assume that the multipoles are located at the origin of a spherical coordinate system (r,θ,ϕ) defined in terms of the cartesian system (x,y,z) by $x = r\sin\theta\cos\phi$, $y = r\sin\theta\sin\phi$, $z = r\cos\theta$. A consequence of this assumption is that the free-space Green's function $G = e^{ikr}/4\pi r$ which appears in expressions (38) and (39) is a function of the radial coordinate r only.

From expressions (38) and (39) we see that the potentials of the electric-dipole part of the source are

$$\phi_{\text{elec.dip.}} = -\frac{1}{\epsilon}\mathbf{p}\cdot\nabla G \tag{41}$$

$$\mathbf{A}_{\text{elec.dip.}} = -i\omega\mu\mathbf{p}G \tag{42}$$

Applying relations (40) to these potentials and using the identities $\nabla(\mathbf{p}\cdot\nabla G) = (\mathbf{p}\cdot\nabla)\nabla G$ and $\nabla\times(\mathbf{p}G) = -\mathbf{p}\times\nabla G$, which follow from vector analysis and the constancy of $\mathbf{p}$, we obtain the electric and magnetic field of the electric dipole:

$$\mathbf{E}_{\text{elec.dip.}} = \frac{1}{\epsilon}[(\mathbf{p}\cdot\nabla)\nabla G + k^2\mathbf{p}G] \tag{43}$$

$$\mathbf{H}_{\text{elec.dip.}} = i\omega\mathbf{p}\times\nabla G \tag{44}$$

Since the gradient operator in spherical coordinates is

$$\nabla = \mathbf{e}_r\frac{\partial}{\partial r} + \mathbf{e}_r\frac{1}{r}\frac{\partial}{\partial\theta} + \mathbf{e}_\phi\frac{1}{r\sin\theta}\frac{\partial}{\partial\phi} \tag{45}$$

the vector ∇G which appears in Eqs. (43) and (44) is given by

$$\nabla G = \nabla\frac{e^{ikr}}{4\pi r} = \mathbf{e}_r\left(ik - \frac{1}{r}\right)G \tag{46}$$

Moreover, in spherical coordinates, $\mathbf{p}$ has the form

$$\mathbf{p} = \mathbf{e}_r p_r + \mathbf{e}_\theta p_\theta + \mathbf{e}_\phi p_\phi \tag{47}$$

where $p = \sqrt{\mathbf{p}\cdot\mathbf{p}} = \sqrt{p_r^2 + p_\theta^2 + p_\phi^2}$ is the strength of the electric dipole. Hence the scalar product of $\mathbf{p}$ and ∇ yields

$$\mathbf{p}\cdot\nabla = p_r\frac{\partial}{\partial r} + p_\theta\frac{1}{r}\frac{\partial}{\partial\theta} + p_\phi\frac{1}{r\sin\theta}\frac{\partial}{\partial\phi} \tag{48}$$

When this operator acts on the vector (46) and it is recalled that $(\partial/\partial r)\mathbf{e}_r = 0$, $(\partial/\partial\theta)\mathbf{e}_r = \mathbf{e}_\theta$, and $(\partial/\partial\phi)\mathbf{e}_r = \mathbf{e}_\phi\sin\theta$, we get

$$(\mathbf{p}\cdot\nabla)\nabla G = \mathbf{e}_r p_r\left(-k^2 - \frac{2ik}{r} + \frac{2}{r^2}\right)G + \mathbf{e}_\theta p_\theta\left(\frac{ik}{r} - \frac{1}{r^2}\right)G + \mathbf{e}_\phi p_\phi\left(\frac{ik}{r} - \frac{1}{r^2}\right)G \tag{49}$$

Using this result and expression (47), we easily obtain from Eq. (43) the spherical component of $\mathbf{E}_{\text{elec.dip.}}$ in terms of the spherical components of **p**. Thus

$$(E_r)_{\text{elec.dip.}} = \frac{1}{\epsilon} p_r \left(-\frac{2ik}{r} + \frac{2}{r^2} \right) G \tag{50}$$

$$(E_\theta)_{\text{elec.dip.}} = \frac{1}{\epsilon} p_\theta \left(\frac{ik}{r} - \frac{1}{r^2} + k^2 \right) G \tag{51}$$

$$(E_\phi)_{\text{elec.dip.}} = \frac{1}{\epsilon} p_\phi \left(\frac{ik}{r} - \frac{1}{r^2} + k^2 \right) G \tag{52}$$

In cartesian coordinates, **p** has the form

$$\mathbf{p} = \mathbf{e}_x p_x + \mathbf{e}_y p_y + \mathbf{e}_z p_z \tag{53}$$

Scalarly multiplying this expression by $\mathbf{e}_r$, $\mathbf{e}_\theta$, $\mathbf{e}_\phi$ in succession, noting that

$$p_r = \mathbf{p} \cdot \mathbf{e}_r \qquad p_\theta = \mathbf{p} \cdot \mathbf{e}_\theta \qquad p_\phi = \mathbf{p} \cdot \mathbf{e}_\phi$$

and recalling that

$$\mathbf{e}_r \cdot \mathbf{e}_x = \sin\theta\cos\phi \qquad \mathbf{e}_r \cdot \mathbf{e}_y = \sin\theta\sin\phi \qquad \mathbf{e}_r \cdot \mathbf{e}_z = \cos\theta$$

$$\mathbf{e}_\theta \cdot \mathbf{e}_x = \cos\theta\cos\phi \qquad \mathbf{e}_\theta \cdot \mathbf{e}_y = \cos\theta\sin\phi \qquad \mathbf{e}_\theta \cdot \mathbf{e}_z = -\sin\theta$$

$$\mathbf{e}_\phi \cdot \mathbf{e}_x = -\sin\phi \qquad \mathbf{e}_\phi \cdot \mathbf{e}_y = \cos\phi$$

we get the following connection between the cartesian and spherical components of **p**:

$$p_r = p_x \sin\theta\cos\phi + p_y \sin\theta\sin\phi + p_z\cos\theta \tag{54}$$

$$p_\theta = p_x \cos\theta\cos\phi + p_y \cos\theta\sin\phi - p_z\sin\theta \tag{55}$$

$$p_\phi = -p_x \sin\phi + p_y\cos\phi \tag{56}$$

Substituting these expressions into Eqs. (50) through (52), we find that the spherical components of $\mathbf{E}_{\text{elec.dip.}}$ in terms of the cartesian components of **p** are given by

$$(E_r)_{\text{elec.dip.}} = \frac{1}{\epsilon}(p_x \sin\theta\cos\phi + p_y \sin\theta\sin\phi + p_z\cos\theta) \times\left(-\frac{2ik}{r}+\frac{2}{r^2}\right)G \quad (57)$$

$$(E_\theta)_{\text{elec.dip.}} = \frac{1}{\epsilon}(p_x \cos\theta\cos\phi + p_y \cos\theta\sin\phi - p_z\sin\theta) \times\left(\frac{ik}{r}-\frac{1}{r^2}+k^2\right)G \quad (58)$$

$$(E_\phi)_{\text{elec.dip.}} = \frac{1}{\epsilon}(-p_x \sin\phi + p_y\cos\phi)\left(\frac{ik}{r}-\frac{1}{r^2}+k^2\right)G \quad (59)$$

To find the spherical components of $\mathbf{H}_{\text{elec.dip.}}$ in terms of the spherical components of **p**, we substitute expressions (46) and (47) into Eq. (44) and thus obtain

$$(H_r)_{\text{elec.dip.}} = 0 \quad (60)$$

$$(H_\theta)_{\text{elec.dip.}} = i\omega p_\phi\left(ik-\frac{1}{r}\right)G \quad (61)$$

$$(H_\phi)_{\text{elec.dip.}} = -i\omega p_\theta\left(ik-\frac{1}{r}\right)G \quad (62)$$

With the aid of Eqs. (55) and (56) we can express these spherical components of $(\mathbf{H})_{\text{elec.dip.}}$ in terms of the cartesian components of **p**:

$$(H_\theta)_{\text{elec.dip.}} = i\omega(-p_x\sin\phi + p_y\cos\phi)\left(ik-\frac{1}{r}\right)G \quad (63)$$

$$(H_\phi)_{\text{elec.dip.}} = -i\omega(p_x\cos\theta\cos\phi + p_y\cos\theta\sin\phi - p_z\sin\theta) \times\left(ik-\frac{1}{r}\right)G \quad (64)$$

Since $(H_r)_{\text{elec.dip.}}$ is identically zero and $(E_r)_{\text{elec.dip.}}$ is not, the radiation field of the electric dipole is an E wave or, equivalently, a TM wave. But as kr is increased, $(E_r)_{\text{elec.dip.}}$ becomes negligibly small compared to $(E_\theta)_{\text{elec.dip.}}$ and $(E_\phi)_{\text{elec.dip.}}$, and therefore in the far zone the radiation field has the structure of a *TEM* wave, and the simple relation $\mathbf{e}_r \times (\mathbf{E})_{\text{elec.dip.}} = \sqrt{\frac{\mu}{\epsilon}}\,(\mathbf{H})_{\text{elec.dip.}}$ is valid there. As kr is decreased, the magnetic field becomes negligible compared to the electric field, and this electric field approaches the electric field of an electrostatic dipole.

As can be seen from expressions (38) and (39), the potentials of the magnetic dipole are

$$\phi_{\text{mag.dip.}} = 0 \tag{65}$$

$$\mathbf{A}_{\text{mag.dip.}} = -\mu\mathbf{m} \times \nabla G \tag{66}$$

When substituted into the second of relations (40), this vector potential yields the magnetic field of the magnetic dipole. That is,

$$\mathbf{H}_{\text{mag.dip.}} = -\nabla \times (\mathbf{m} \times \nabla G) \tag{67}$$

Since $\mathbf{m}$ is a constant vector, it follows from vector analysis that $\nabla \times (\mathbf{m} \times \nabla G) = \mathbf{m}\nabla^2 G - (\mathbf{m} \cdot \nabla)\nabla G$. But $\nabla^2 G = -k^2 G$ for $r > 0$. Hence the magnetic field (67) of the magnetic dipole may be written alternatively as

$$\mathbf{H}_{\text{mag.dip.}} = -[(\mathbf{m} \cdot \nabla)\nabla G + k^2\mathbf{m}G] \tag{68}$$

The electric field of the magnetic dipole is found by substituting expressions (65) and (66) into the first of relations (40). Thus

$$\mathbf{E}_{\text{mag.dip.}} = -i\omega\mu\mathbf{m} \times \nabla G \tag{69}$$

On comparing Eq. (43) with Eq. (68) and Eq. (44) with Eq. (69), we see that the electromagnetic field of an electric dipole, except for certain multiplicative factors, is formally equivalent to the electromagnetic field of a magnetic dipole, with electric and magnetic fields interchanged. Hence, we can obtain the components of the magnetic dipole by simply applying a duality transformation to the already obtained field components of the electric dipole. Since the radiation field of an electric dipole is an E wave and the dual of an E wave is an H wave, the radiation field of a magnetic dipole thus must be an H wave or, equivalently, a *TE* wave.

According to expressions (38) and (39), the potentials of the electric quadrupole are

$$\phi_{\text{elec.quad.}} = \frac{1}{2\epsilon}\mathbf{Q}:\nabla\nabla G \tag{70}$$

$$\mathbf{A}_{\text{elec.quad.}} = \frac{i\omega\mu}{2}\mathbf{Q}\cdot\nabla G \tag{71}$$

With the aid of relations (40), these potentials yield the following expressions for the electric and magnetic fields of the electric quadrupole:

$$\mathbf{E}_{\text{elec.quad.}} = -\frac{1}{2\epsilon}[\nabla(\mathbf{Q}:\nabla\nabla G) + k^2(\mathbf{Q}\cdot\nabla G)] \tag{72}$$

$$\mathbf{H}_{\text{elec.quad.}} = \frac{i\omega}{2}\nabla\times(\mathbf{Q}\cdot\nabla G) \tag{73}$$

First let us find the components of $\mathbf{H}_{\text{elec.quad.}}$. In spherical coordinates, $\mathbf{Q}$ can be written as

$$\mathbf{Q} = \mathbf{e}_r\mathbf{e}_r Q_{rr} + \mathbf{e}_r\mathbf{e}_\theta Q_{r\theta} + \mathbf{e}_r\mathbf{e}_\phi Q_{r\phi} + \mathbf{e}_\theta\mathbf{e}_r Q_{\theta r} + \mathbf{e}_\theta\mathbf{e}_\theta Q_{\theta\theta} + \mathbf{e}_\theta\mathbf{e}_\phi Q_{\theta\phi} + \mathbf{e}_\phi\mathbf{e}_r Q_{\phi r} + \mathbf{e}_\phi\mathbf{e}_\theta Q_{\phi\theta} + \mathbf{e}_\phi\mathbf{e}_\phi Q_{\phi\phi} \tag{74}$$

Scalarly postmultiplying this expression by $\nabla G = \mathbf{e}_r f$, where f is a shorthand for $(ik - 1/r)G$, we get the vector

$$\mathbf{Q}\cdot\nabla G = (\mathbf{e}_r Q_{rr} + \mathbf{e}_\theta Q_{\theta r} + \mathbf{e}_\phi Q_{\phi r})f \tag{75}$$

which, when substituted into Eq. (73), yields

$$(H_r)_{\text{elec.quad.}} = \frac{i\omega}{2}\left[\frac{\partial}{\partial\theta}(Q_{\phi r}\sin\theta) - \frac{\partial}{\partial\phi}Q_{\theta r}\right]\frac{f}{r\sin\theta} \tag{76}$$

$$(H_\theta)_{\text{elec.quad.}} = \frac{i\omega}{2}\left[\frac{f}{r\sin\theta}\frac{\partial}{\partial\phi}Q_{rr} - Q_{\phi r}\frac{1}{r}\frac{\partial}{\partial r}(rf)\right] \tag{77}$$

$$(H_\phi)_{\text{elec.quad.}} = \frac{i\omega}{2}\left[Q_{\theta r}\frac{1}{r}\frac{\partial}{\partial r}(rf) - \frac{f}{r}\frac{\partial}{\partial\theta}Q_{rr}\right] \tag{78}$$

Since

$$\frac{1}{r}\frac{\partial}{\partial r}(rf) = \frac{1}{r}\frac{\partial}{\partial r}\left[(ikr - 1)\frac{e^{ikr}}{4\pi r}\right] \rightarrow -k^2\frac{e^{ikr}}{4\pi r} = -k^2 G \tag{79}$$

as $r \rightarrow \infty$, the only parts of the magnetic field components that survive in the far zone are

$$(H_\theta)_{\text{elec.quad.}} = \frac{i\omega k^2}{2} Q_{\phi r} G \tag{80}$$

$$(H_\phi)_{\text{elec.quad.}} = -\frac{i\omega k^2}{2} Q_{\theta r} G \tag{81}$$

To represent these far-zone field components in terms of the cartesian components of **Q**, we note that in cartesian coordinates **Q** has the form

$$\begin{aligned}\mathbf{Q} = \mathbf{e}_x\mathbf{e}_x Q_{xx} + \mathbf{e}_x\mathbf{e}_y Q_{xy} + \mathbf{e}_x\mathbf{e}_z Q_{xz} + \mathbf{e}_y\mathbf{e}_x Q_{yx} + \mathbf{e}_y\mathbf{e}_y Q_{yy} + \mathbf{e}_y\mathbf{e}_z Q_{yz} \\ + \mathbf{e}_z\mathbf{e}_x Q_{zx} + \mathbf{e}_z\mathbf{e}_y Q_{zy} + \mathbf{e}_z\mathbf{e}_z Q_{zz}\end{aligned} \tag{82}$$

which when premultiplied by $\mathbf{e}_\theta$, $\mathbf{e}_\phi$ and postmultiplied by $\mathbf{e}_r$ yields

$$\begin{aligned}Q_{\theta r} = \mathbf{e}_\theta \cdot \mathbf{Q} \cdot \mathbf{e}_r = Q_{xx} \cos\theta \sin\theta \cos^2\phi + Q_{yy} \cos\theta \sin\theta \sin^2\phi \\ - Q_{zz} \sin\theta \cos\theta + (Q_{xy} + Q_{yx}) \cos\theta \sin\theta \cos\phi \sin\phi \\ + Q_{xz} \cos^2\theta \cos\phi - Q_{zx} \sin^2\theta \cos\phi + Q_{yz} \cos^2\theta \sin\phi - Q_{zy} \sin^2\theta \sin\phi\end{aligned}$$

and

$$\begin{aligned}Q_{\phi r} = \mathbf{e}_\phi \cdot \mathbf{Q} \cdot \mathbf{e}_r = (Q_{yy} - Q_{xx}) \sin\theta \sin\phi \cos\phi - Q_{xy} \sin\theta \sin^2\phi \\ + Q_{yx} \sin\theta \cos^2\phi - Q_{xz} \sin\phi \cos\theta + Q_{yz} \cos\phi \cos\theta\end{aligned}$$

Invoking the symmetry of **Q** and using some simple trigonometric identities, we reduce these results to

$$\begin{aligned}Q_{\theta r} = \tfrac{1}{2} \sin 2\theta (Q_{xx} \cos^2\phi + Q_{yy} \sin^2\phi - Q_{zz} + Q_{xy} \sin 2\phi) \\ + \cos 2\theta (Q_{xz} \cos\phi + Q_{yz} \sin\phi)\end{aligned} \tag{83}$$

$$\begin{aligned}Q_{\phi r} = \tfrac{1}{2}(Q_{yy} - Q_{xx}) \sin\theta \sin 2\phi + Q_{yx} \sin\theta \cos 2\phi \\ - Q_{xz} \sin\phi \cos\theta + Q_{yz} \cos\phi \cos\theta\end{aligned} \tag{84}$$

Substituting Eqs. (84) and (83) into Eqs. (80) and (81), we obtain the following expressions for the spherical components of the far-zone magnetic field of an electric quadrupole in terms of the cartesian components of the quadrupole moment:

$$(H_\theta)_{\text{elec.quad.}} = \frac{i\omega k^2}{2}\frac{e^{ikr}}{4\pi r}[\tfrac{1}{2}(Q_{yy} - Q_{xx}) \sin\theta \sin 2\phi$$
$$+ Q_{yx} \sin\theta \cos 2\phi - Q_{xz} \sin\phi \cos\theta + Q_{yz} \cos\phi \cos\theta] \quad (85)$$
$$(H_\phi)_{\text{elec.quad.}} = -\frac{i\omega k^2}{2}\frac{e^{ikr}}{4\pi r}[\tfrac{1}{2} \sin 2\theta(Q_{xx} \cos^2\phi + Q_{yy} \sin^2\phi$$
$$- Q_{zz} + Q_{xy} \sin 2\phi) + \cos 2\theta(Q_{xz} \cos\phi + Q_{yz} \sin\phi)] \quad (86)$$

In the far zone the relation $\mathbf{e}_r \times \mathbf{E}_{\text{elec.quad.}} = \sqrt{\frac{\mu}{\epsilon}}\,\mathbf{H}_{\text{elec.quad.}}$ is valid. Consequently the spherical components of the far-zone electric field of the electric quadrupole are derivable from expressions (85) and (86) by use of the following simple connections:

$$(E_\theta)_{\text{elec.quad.}} = \sqrt{\frac{\mu}{\epsilon}}\,(H_\phi)_{\text{elec.quad.}} \quad (87)$$

$$(E_\phi)_{\text{elec.quad.}} = -\sqrt{\frac{\mu}{\epsilon}}\,(H_\theta)_{\text{elec.quad.}} \quad (88)$$

Alternatively one may calculate the far-zone components of $\mathbf{E}_{\text{elec.quad.}}$ directly from Eq. (72). We have already calculated the quantity $\mathbf{Q} \cdot \nabla G$ which appears in the second term of Eq. (72). The result of this calculation is shown by Eq. (75). Hence, the only quantity we now must calculate is $\nabla(\mathbf{Q}:\nabla\nabla G)$ for $r \to \infty$. By definition of the double scalar product, we have

$$\mathbf{Q}:\nabla\nabla G = \sum_{i,j} Q_{ij}(\mathbf{e}_i \cdot \nabla)(\mathbf{e}_j \cdot \nabla)G \quad (89)$$

where $\mathbf{e}_i$ denotes the unit vectors in spherical coordinates. Since G is a function of r only, this definition yields the expression

$$\mathbf{Q}:\nabla\nabla G = Q_{rr}\frac{\partial^2}{\partial r^2} G \quad (90)$$

which, in the far zone, reduces to

$$\mathbf{Q}:\nabla\nabla G = -k^2 Q_{rr} G \qquad (r \to \infty) \tag{91}$$

Taking the gradient of this quantity and keeping only its far-zone term, we get

$$\nabla(\mathbf{Q}:\nabla\nabla G) = -ik^3 Q_{rr} G \mathbf{e}_r \qquad (r \to \infty) \tag{92}$$

Substituting Eqs. (75) and (92) into Eq. (72), we see that in the far zone $(E_r)_{\text{elec.quad.}}$ disappears and the other two components of $\mathbf{E}_{\text{elec.quad.}}$ are given by

$$(E_\theta)_{\text{elec.quad.}} = -\frac{ik^3}{2\epsilon} Q_{\theta r} G \tag{93}$$

$$(E_\phi)_{\text{elec.quad.}} = -\frac{ik^3}{2\epsilon} Q_{\phi r} G \tag{94}$$

With the aid of Eqs. (80) and (81) it is clear that this result agrees with Eqs. (87) and (88).

4.4 Expansion of Radiation Field in Spherical Waves

There is an alternative type of multipole expansion which in certain instances is more natural than the one based on the Taylor series expansion. In this section we shall construct such an expansion by first developing the radiation in spherical E and H waves, then defining the E waves as electric multipoles and the H waves as magnetic multipoles, and finally calculating the expansion coefficients through Bouwkamp and Casimir's method.[1]

Outside the bounded region V_0, which completely contains the monochromatic source currents, the electric and magnetic fields can be con-

[1] C. J. Bouwkamp and H. B. G. Casimir, On Multipole Expansions in the Theory of Electromagnetic Radiation, *Physica*, **20:** 539 (1954).

veniently derived from two scalar functions by use of the expressions

$$\mathbf{E} = \nabla \times \nabla \times (\mathbf{r}v) + i\omega\mu\nabla \times (\mathbf{r}u) \tag{95}$$

$$\mathbf{H} = \nabla \times \nabla \times (\mathbf{r}u) - i\omega\epsilon\nabla \times (\mathbf{r}v) \tag{96}$$

The two scalar functions u and v are the Debye potentials[1] which satisfy the scalar Helmholtz equation

$$(\nabla^2 + k^2)\psi = 0 \tag{97}$$

and obey the Sommerfeld radiation condition. Such a representation of an electromagnetic field in terms of the Debye potentials is quite general. Indeed, it has been proved[2] that every electromagnetic field in a source-free region between two concentric spheres can be represented by the Debye potentials; the proof rests on Hodge's decomposition theorem for vector fields defined on a sphere.[3]

We choose a spherical coordinate system (r,θ,ϕ) with center somewhere within V_0. Then the acceptable solutions of Eq. (97) are the spherical wave functions

$$\psi_n{}^m(r,\theta,\phi) = h_n(kr)Y_n{}^m(\theta,\phi) \qquad (n \geq 0,\ m = 0, \pm 1, \ldots, \pm n) \tag{98}$$

The radial functions $h_n(kr)$ are the spherical Hankel functions of the first kind, which satisfy the differential equation

$$r^2\frac{d^2h_n}{dr^2} + 2r\frac{dh_n}{dr} + [k^2r^2 - n(n+1)]h_n = 0 \tag{99}$$

and obey the radiation condition. They are related to the fractional-order cylindrical Hankel functions of the first kind by

$$h_n(kr) = (\pi/2kr)^{1/2}\,H_{n+1/2}{}^{(1)}(kr)$$

[1] P. Debye, Dissertation, Munich, 1908; also, Der Lichtdruck auf Kugeln von beliebigen Material, *Ann. Phys.*, **30:** 57 (1909).

[2] C. H. Wilcox, Debye Potentials, *J. Math. Mech.*, **6:** 167 (1957).

[3] P. Bidal and G. de Rham, Les formes differentielles harmoniques, *Commentarii Mathematici Helvetici*, **19:** 1 (1956).

The fact that h_n can be expressed in terms of the exponential function is sometimes useful; for example[1]

$$h_0(kr) = -\frac{i\,e^{ikr}}{kr} \qquad h_1(kr) = -\frac{e^{ikr}}{kr} - \frac{i\,e^{ikr}}{(kr)^2}$$
$$h_2(kr) = \frac{i\,e^{ikr}}{kr} - \frac{3e^{ikr}}{(kr)^2} - \frac{3i\,e^{ikr}}{(kr)^3} \qquad \text{etc.} \tag{100}$$

The angular functions $Y_n^m(\theta,\phi)$ are the surface spherical harmonics of degree n and order m. They constitute a complete set of orthogonal functions on the surface of a sphere. Displaying explicitly the normalization constant, we write

$$Y_n^m(\theta,\phi) = \left[(2n+1)\frac{(n-m)!}{(n+m)!}\right]^{1/2} P_n^m(\cos\theta)e^{im\phi} \qquad (-n \le m \le n) \tag{101}$$

where $P_n^m(\cos\theta)$ are the associated Legendre polynomials of degree n and order m; then in view of

$$\int_0^\pi [P_n^m(\cos\theta)]^2 \sin\theta\, d\theta = \frac{2}{2n+1}\frac{(n+m)!}{(n-m)!} \qquad (-n \le m \le n) \tag{102}$$

we see that this choice of normalization constant leads to the orthogonality relations

$$\frac{\int_0^{2\pi}\int_0^\pi Y_{n'}^{m'} Y_n^{m*} \sin\theta\, d\theta\, d\phi}{\int_0^{2\pi}\int_0^\pi \sin\theta\, d\theta\, d\phi} = \delta_{nn'}\delta_{mm'} \tag{103}$$

where $\delta_{ij} = 1$ for $i = j$ and 0 for $i \neq j$. Here the quantity Y_n^{m*} is the conjugate complex of Y_n^m and is simply related to Y_n^{-m} as follows:

$$Y_n^{m*} = (-1)^m Y_n^{-m} \tag{104}$$

[1] See P. M. Morse, "Vibration and Sound," 2d ed., pp. 316–317, McGraw-Hill Book Company, New York, 1948.

This relation clearly follows from definition (101) when we recall

$$P_n^{-m}(\cos\theta) = (-1)^m \frac{(n-m)!}{(n+m)!} P_n^m(\cos\theta) \tag{105}$$

The Debye potentials are linear superpositions of these spherical wave functions (98). That is,

$$v(r,\theta,\phi) = \sum_{n=0}^{\infty} \sum_{m=-n}^{m=n} a_{nm}\psi_n^m \tag{106}$$

$$u(r,\theta,\phi) = \sum_{n=0}^{\infty} \sum_{m=-n}^{m=n} b_{nm}\psi_n^m \tag{107}$$

The expansion coefficients a_{nm}, b_{nm} could be calculated from a knowledge of u and v on the surface of a sphere of radius $r = r_0$ by using the property that the functions ψ_n^m are orthogonal over the surface of the sphere, viz.,

$$\int_0^{2\pi}\int_0^{\pi} \psi_n^m(\psi_{n'}^{m'})^* \sin\theta \, d\theta \, d\phi = 4\pi h_n(kr_0)h_{n'}^*(kr_0)\delta_{nn'}\delta_{mm'} \tag{108}$$

However, we shall not determine them in this way. Rather, we shall determine them from the radial components of the electric and magnetic fields, in accord with the method of Bouwkamp and Casimir.[1]

Substituting expressions (106) and (107) for the Debye potentials into representations (95) and (96), we get the following expansion of the electromagnetic field in spherical wave functions:

$$\mathbf{E} = \sum_{n,m} a_{nm}\nabla \times \nabla \times (\mathbf{r}\psi_n^m) + i\omega\mu \sum_{n,m} b_{nm}\nabla \times (\mathbf{r}\psi_n^m) \tag{109}$$

$$\mathbf{H} = \sum_{n,m} b_{nm}\nabla \times \nabla \times (\mathbf{r}\psi_n^m) - i\omega\epsilon \sum_{n,m} a_{nm}\nabla \times (\mathbf{r}\psi_n^m) \tag{110}$$

We are free to consider this electromagnetic field as a superposition of two electromagnetic fields, one being an E type field ($E_r \neq 0$, $H_r = 0$) and the other an H type field ($H_r \neq 0$, $E_r = 0$). Accordingly we

[1] C. J. Bouwkamp and H. B. G. Casimir, On Multipole Expansions in the Theory of Electromagnetic Radiation, *Physica*, **20:** 539 (1954). Also, H. B. G. Casimir, A Note on Multipole Radiation, *Helv. Phys. Acta*, **33:** 849 (1960).

decompose the electromagnetic field **E**, **H** as follows:

$$\mathbf{E} = \mathbf{E}' + \mathbf{E}'' \tag{111}$$

$$\mathbf{H} = \mathbf{H}'' + \mathbf{H}' \tag{112}$$

where $\mathbf{E}'$, $\mathbf{H}'$ denote the E type field and $\mathbf{E}''$, $\mathbf{H}''$ the H type field. Comparing Eqs. (109) and (110) with Eqs. (111) and (112) respectively, we see that the E type field is given by

$$\mathbf{E}' = \sum_{n,m} a_{nm} \nabla \times \nabla \times (\mathbf{r}\psi_n{}^m) \tag{113}$$

$$\mathbf{H}' = -i\omega\epsilon \sum_{n,m} a_{nm} \nabla \times (\mathbf{r}\psi_n{}^m) \tag{114}$$

and the H type field by

$$\mathbf{E}'' = i\omega\mu \sum_{n,m} b_{nm} \nabla \times (\mathbf{r}\psi_n{}^m) \tag{115}$$

$$\mathbf{H}'' = \sum_{n,m} b_{nm} \nabla \times \nabla \times (\mathbf{r}\psi_n{}^m) \tag{116}$$

Moreover, if we let

$$\left.\begin{aligned} \mathbf{E}'_{nm} &= \nabla \times \nabla \times (\mathbf{r}\psi_n{}^m) \\ \mathbf{H}'_{nm} &= -i\omega\epsilon \nabla \times (\mathbf{r}\psi_n{}^m) \end{aligned}\right\} \text{electric multipoles of degree } n \text{ and order } m \tag{117, 118}$$

and

$$\left.\begin{aligned} \mathbf{E}''_{nm} &= i\omega\mu \nabla \times (\mathbf{r}\psi_n{}^m) \\ \mathbf{H}''_{nm} &= \nabla \times \nabla \times (\mathbf{r}\psi_n{}^m) \end{aligned}\right\} \text{magnetic multipoles of degree } n \text{ and order } m \tag{119, 120}$$

then Eqs. (113) through (116) become

$$\mathbf{E}' = \sum_{n,m} a_{nm} \mathbf{E}'_{nm} \tag{121}$$

$$\mathbf{H}' = \sum_{n,m} a_{nm} \mathbf{H}'_{nm} \tag{122}$$

$$\mathbf{E}'' = \sum_{n,m} b_{nm} \mathbf{E}''_{nm} \tag{123}$$

$$\mathbf{H}'' = \sum_{n,m} b_{nm} \mathbf{H}''_{nm} \tag{124}$$

We define $\mathbf{E}'_{nm}$, $\mathbf{H}'_{nm}$ to be the electromagnetic field of the electric multipole of degree n and order m, and $\mathbf{E}''_{nm}$, $\mathbf{H}''_{nm}$ to be the electromagnetic field of the magnetic multipole of degree n and order m, so that expressions (121) through (124) constitute the multipole expansion of the electromagnetic field. Thus a superposition of the terms $n = 1$, $m = 0, \pm 1$ yields a dipolar field, and a superposition of the terms $n = 2$, $m = 0, \pm 1, \pm 2$ yields a quadrupolar field, and so forth.

As yet the expansion coefficients have not been fixed; before we start to calculate them, let us deduce the spherical components of the multipole fields. To find the spherical components of $\mathbf{E}'_{nm}$, $\mathbf{H}'_{nm}$ and $\mathbf{E}''_{nm}$, $\mathbf{H}''_{nm}$, we make use of the following relations. We note that

$$\mathbf{r} \cdot \nabla \times \nabla \times (\mathbf{r}\psi_n{}^m) = \left(r^2 \frac{\partial^2}{\partial r^2} + 2r \frac{\partial}{\partial r} + k^2 r^2\right) \psi_n{}^m$$

$$= n(n+1)\psi_n{}^m \quad (125)$$

where the second equality follows from Eq. (99). We also note that

$$\mathbf{r} \cdot \nabla \times (\mathbf{r}\psi_n{}^m) = \mathbf{r} \cdot (\nabla\psi_n{}^m \times \mathbf{r}) = 0 \quad (126)$$

Denoting the unit vectors in the θ and ϕ directions by $\mathbf{e}_\theta$ and $\mathbf{e}_\phi$ respectively, we obtain by vector analysis the angular components of $\nabla \times (\mathbf{r}\psi_n{}^m)$ and $\nabla \times \nabla \times (\mathbf{r}\psi_n{}^m)$:

$$\mathbf{e}_\theta \cdot \nabla \times \nabla \times (\mathbf{r}\psi_n{}^m) = \frac{1}{r} \frac{\partial}{\partial r} \frac{\partial}{\partial \theta} (r\psi_n{}^m) \quad (127)$$

$$\mathbf{e}_\phi \cdot \nabla \times \nabla \times (\mathbf{r}\psi_n{}^m) = \frac{1}{r \sin\theta} \frac{\partial}{\partial r} \frac{\partial}{\partial \phi} (r\psi_n{}^m) \quad (128)$$

$$\mathbf{e}_\theta \cdot \nabla \times (\mathbf{r}\psi_n{}^m) = \frac{1}{\sin\theta} \frac{\partial}{\partial \phi} \psi_n{}^m \quad (129)$$

$$\mathbf{e}_\phi \cdot \nabla \times (\mathbf{r}\psi_n{}^m) = -\frac{\partial}{\partial \theta} \psi_n{}^m \quad (130)$$

With the aid of relations (125) through (130) we see that the spherical components of the multipole fields (117) through (120) are given by

$$\left.\begin{aligned}
(E'_{nm})_r &= \frac{n(n+1)}{r} h_n Y_n{}^m \\
(E'_{nm})_\theta &= \frac{1}{r}\frac{d}{dr}(rh_n)\frac{\partial}{\partial\theta} Y_n{}^m \\
(E'_{nm})_\phi &= \frac{im}{r\sin\theta}\frac{d}{dr}(rh_n) Y_n{}^m \\
(H'_{nm})_r &= 0 \\
(H'_{nm})_\theta &= \frac{m\omega\epsilon}{\sin\theta} h_n Y_n{}^m \\
(H'_{nm})_\phi &= i\omega\epsilon h_n \frac{\partial}{\partial\theta} Y_n{}^m
\end{aligned}\right\}
\begin{array}{l}\text{spherical components of}\\ \text{electric multipole field}\\ \text{of degree } n \text{ and order } m\\ (n \geq 0,\ -n \leq m \leq n)\end{array} \tag{131}$$

and

$$\left.\begin{aligned}
(H''_{nm})_r &= \frac{n(n+1)}{r} h_n Y_n{}^m \\
(H''_{nm})_\theta &= \frac{1}{r}\frac{d}{dr}(rh_n)\frac{\partial}{\partial\theta} Y_n{}^m \\
(H''_{nm})_\phi &= \frac{im}{r\sin\theta}\frac{d}{dr}(rh_n) Y_n{}^m \\
(E''_{nm})_r &= 0 \\
(E''_{nm})_\theta &= -\frac{m\omega\mu}{\sin\theta} h_n Y_n{}^m \\
(E''_{nm})_\phi &= -i\omega\mu h_n \frac{\partial}{\partial\theta} Y_n{}^m
\end{aligned}\right\}
\begin{array}{l}\text{spherical components of}\\ \text{magnetic multipole field}\\ \text{of degree } n \text{ and order } m\\ (n \geq 0,\ -n \leq m \leq n)\end{array} \tag{132}$$

With the aid of relations (125) and (126) it follows from Eqs. (109) and (110) that $\mathbf{r}\cdot\mathbf{E}$ and $\mathbf{r}\cdot\mathbf{H}$ can be written as[1]

$$\mathbf{r}\cdot\mathbf{E} = \sum_{n,m} a_{nm} n(n+1)\psi_n{}^m \tag{133}$$

$$\mathbf{r}\cdot\mathbf{H} = \sum_{n,m} b_{nm} n(n+1)\psi_n{}^m \tag{134}$$

By virtue of the orthogonality of the functions $\psi_n{}^m$ over the surface of a sphere, it is obvious from expansions (133) and (134) that the coeffi-

[1] Comparing these expressions for $\mathbf{r}\cdot\mathbf{E}$ and $\mathbf{r}\cdot\mathbf{H}$ with expressions (106) and (107) for v and u we see that apart from the factor $n(n+1)$ they are respectively the same.

cients a_{nm} and b_{nm} can be determined from a knowledge of the scalar functions $\mathbf{r} \cdot \mathbf{E}$ and $\mathbf{r} \cdot \mathbf{H}$ over the surface of a sphere. Accordingly we now shall find the expansion coefficients a_{nm}, b_{nm} from the currents within V_0 by first calculating $\mathbf{r} \cdot \mathbf{E}$ and $\mathbf{r} \cdot \mathbf{H}$ in terms of the currents and then incorporating these results with expansions (133) and (134). The validity of this procedure is assured by the theorem[1] that any electromagnetic field in the empty space between two concentric spheres is completely determined by the radial components E_r, H_r.

Our task now is to obtain $\mathbf{r} \cdot \mathbf{E}$ and $\mathbf{r} \cdot \mathbf{H}$ in terms of the currents within V_0. We recall (see Sec. 1.1) that the $\mathbf{E}$ and $\mathbf{H}$ produced by a monochromatic $\mathbf{J}$ must satisfy the Helmholtz equations

$$\nabla \times \nabla \times \mathbf{H} - k^2\mathbf{H} = \nabla \times \mathbf{J} \tag{135}$$

$$\nabla \times \nabla \times \mathbf{E} - k^2\mathbf{E} = i\omega\mu\mathbf{J} \tag{136}$$

When these equations are scalarly multiplied by the position vector $\mathbf{r}$, we get the relations

$$\mathbf{r} \cdot \nabla \times \nabla \times \mathbf{H} - k^2\mathbf{r} \cdot \mathbf{H} = \mathbf{r} \cdot \nabla \times \mathbf{J} \tag{137}$$

$$\mathbf{r} \cdot \nabla \times \nabla \times \mathbf{E} - k^2\mathbf{r} \cdot \mathbf{E} = i\omega\mu\mathbf{r} \cdot \mathbf{J} \tag{138}$$

which by vector analysis reduce[2] to

$$(\nabla^2 + k^2)(\mathbf{r} \cdot \mathbf{H}) = -\mathbf{r} \cdot \nabla \times \mathbf{J} \tag{139}$$

$$(\nabla^2 + k^2)\left(\mathbf{r} \cdot \mathbf{E} + \frac{i}{\omega\epsilon}\mathbf{r} \cdot \mathbf{J}\right) = \frac{1}{i\omega\epsilon}\mathbf{r} \cdot \nabla \times \nabla \times \mathbf{J} \tag{140}$$

In terms of the free-space Green's function

$$G(\mathbf{r},\mathbf{r}') = \frac{e^{ik|\mathbf{r}-\mathbf{r}'|}}{4\pi|\mathbf{r} - \mathbf{r}'|} \tag{141}$$

[1] Bouwkamp and Casimir, *loc. cit.*

[2] We use the vector identity $(\nabla^2 + k^2)(\mathbf{r} \cdot \mathbf{C}) = 2\nabla \cdot \mathbf{C} + \mathbf{r} \cdot \nabla(\nabla \cdot \mathbf{C}) - \mathbf{r} \cdot \nabla \times \nabla \times \mathbf{C} + k^2\mathbf{r} \cdot \mathbf{C}$, where $\mathbf{C}$ is an arbitrary vector field. To obtain Eq. (139) we let $\mathbf{C} = \mathbf{H}$. To obtain Eq. (140) we let $\mathbf{C} = \mathbf{E}$ and $\mathbf{C} = \mathbf{J}$ successively.

the solutions of the scalar Helmholtz equations (139) and (140) which satisfy the Sommerfeld radiation condition are

$$\mathbf{r}\cdot\mathbf{H} = \int_{V_0} G(\mathbf{r},\mathbf{r}')\mathbf{r}'\cdot\nabla'\times\mathbf{J}(\mathbf{r}')dV' \tag{142}$$

$$\mathbf{r}\cdot\mathbf{E} = \frac{1}{i\omega\epsilon}\mathbf{r}\cdot\mathbf{J} - \frac{1}{i\omega\epsilon}\int_{V_0} G(\mathbf{r},\mathbf{r}')\mathbf{r}'\cdot\nabla'\times\nabla'\times\mathbf{J}(\mathbf{r}')dV' \tag{143}$$

These relations are valid for $\mathbf{r}$ inside and outside V_0. For $\mathbf{r}$ outside V_0, the first term on the right side of Eq. (143) is identically zero and we have

$$\mathbf{r}\cdot\mathbf{E} = \frac{i}{\omega\epsilon}\int_{V_0} G(\mathbf{r},\mathbf{r}')\mathbf{r}'\cdot\nabla'\times\nabla'\times\mathbf{J}(\mathbf{r}')dV' \tag{144}$$

It is known that for $r' < r$

$$G(\mathbf{r},\mathbf{r}') = \frac{e^{ik|\mathbf{r}-\mathbf{r}'|}}{4\pi|\mathbf{r}-\mathbf{r}'|} = \frac{ik}{4\pi}\sum_{n=0}^{n=\infty}(2n+1)j_n(kr')h_n(kr)P_n(\cos\gamma) \tag{145}$$

where $j_n(kr') = (\pi/2kr')^{1/2}J_{n+1/2}(kr')$ and $\cos\gamma = \cos\theta\cos\theta' + \sin\theta\sin\theta'\cos(\phi-\phi')$. It is also known that

$$P_n(\cos\gamma) = \sum_{m=-n}^{m=n}(-1)^m P_n^{\ m}(\cos\theta)P_n^{\ -m}(\cos\theta')e^{im(\phi-\phi')} \tag{146}$$

Recalling

$$\begin{aligned}\psi_n^{\ m} &= h_n(kr)Y_n^{\ m}(\theta,\phi)\\ &= \left[(2n+1)\frac{(n-m)!}{(n+m)!}\right]^{1/2} h_n(kr)P_n^{\ m}(\cos\theta)e^{im\phi}\end{aligned} \tag{147}$$

and introducing the functions $\chi_n^{\ m}$, which are defined by

$$\begin{aligned}\chi_n^{\ m} &= j_n(kr)Y_n^{\ m}(\theta,\phi)\\ &= \left[(2n+1)\frac{(n-m)!}{(n+m)!}\right]^{1/2} j_n(kr)P_n^{\ m}(\cos\theta)e^{im\phi}\end{aligned} \tag{148}$$

we find from expansions (145) and (146) that the free-space Green's function can be expressed as follows:

$$G(\mathbf{r},\mathbf{r}') = \frac{e^{ik|\mathbf{r}-\mathbf{r}'|}}{4\pi|\mathbf{r}-\mathbf{r}'|} = \frac{ik}{4\pi}\sum_{n,m}(-1)^m\chi_n{}^{-m}(\mathbf{r}')\psi_n{}^m(\mathbf{r}) \tag{149}$$

Substituting this expression into Eq. (142) we get

$$\mathbf{r}\cdot\mathbf{H} = \frac{ik}{4\pi}\sum_{n,m}(-1)^m\psi_n{}^m(\mathbf{r})\int_{V_0}\chi_n{}^{-m}(\mathbf{r}')\mathbf{r}'\cdot\nabla'\times\mathbf{J}(\mathbf{r}')dV' \tag{150}$$

An integration by parts yields

$$\int_{V_0}\chi_n{}^{-m}(\mathbf{r}')\mathbf{r}'\cdot\nabla'\times\mathbf{J}(\mathbf{r}')dV' = \int_{V_0}\mathbf{J}(\mathbf{r}')\cdot\nabla'\times(\mathbf{r}'\chi_n{}^{-m})dV' \tag{151}$$

and hence the expansion for $\mathbf{r}\cdot\mathbf{H}$ becomes

$$\mathbf{r}\cdot\mathbf{H} = \frac{ik}{4\pi}\sum_{n,m}(-1)^m\psi_n{}^m(\mathbf{r})\int_{V_0}\mathbf{J}(\mathbf{r}')\cdot\nabla'\times(\mathbf{r}'\chi_n{}^{-m})dV' \tag{152}$$

From Eq. (144) it similarly follows that

$$\mathbf{r}\cdot\mathbf{E} = -\frac{1}{4\pi}\sqrt{\frac{\mu}{\epsilon}}\sum_{n,m}(-1)^m\psi_n{}^m(\mathbf{r})\int_{V_0}\mathbf{J}(\mathbf{r}')\cdot\nabla' \times\nabla'\times(\mathbf{r}'\chi_n{}^{-m})dV' \tag{153}$$

Comparing Eq. (152) with Eq. (134) and Eq. (153) with Eq. (133), we finally obtain the desired formulas

$$a_{nm} = -\frac{1}{4\pi}\sqrt{\frac{\mu}{\epsilon}}\frac{(-1)^m}{n(n+1)}\int_{V_0}\mathbf{J}(\mathbf{r}')\cdot\nabla'\times\nabla'\times(\mathbf{r}'\chi_n{}^{-m})dV' \tag{154}$$

$$b_{nm} = \frac{ik}{4\pi}\frac{(-1)^m}{n(n+1)}\int_{V_0}\mathbf{J}(\mathbf{r}')\cdot\nabla'\times(\mathbf{r}'\chi_n{}^{-m})dV' \tag{155}$$

which give a_{nm} and b_{nm} in terms of the current.

From the above analysis we see that a multipole expansion of the electromagnetic field **E**, **H** radiated by a monochromatic current **J** is obtained by developing **E** and **H** in basic multipole fields $\mathbf{E}'_{nm}$, $\mathbf{H}'_{nm}$ and $\mathbf{E}''_{nm}$, $\mathbf{H}''_{nm}$, that is, by writing **E** and **H** in the following form:

$$\mathbf{E} = \sum_{n,m} a_{nm}\mathbf{E}'_{nm} + \sum_{n,m} b_{nm}\mathbf{E}''_{nm} \tag{156}$$

$$\mathbf{H} = \sum_{n,m} a_{nm}\mathbf{H}'_{nm} + \sum_{n,m} b_{nm}\mathbf{H}''_{nm} \tag{157}$$

Here the basic multipole fields are given explicitly in terms of spherical wave functions by the definitions (117) through (120), and the expansion coefficients a_{nm}, b_{nm} (which constitute a decomposition of the known current **J** into electric and magnetic multipoles superposed at the origin of coordinates) are deduced from **J** by evaluating the integrals (154) and (155).

In the far zone ($kr \to \infty$), the basic multipole fields can be expressed most conveniently in terms of the operator

$$\mathbf{L} = \frac{1}{i}\mathbf{r} \times \nabla \tag{158}$$

which in wave mechanics is known as the angular-momentum operator. To show this, we note that the asymptotic form of the spherical Hankel function is

$$h_n(kr) \approx (-i)^{n+1}\frac{e^{ikr}}{kr} \qquad (kr \to \infty) \tag{159}$$

From this form it follows that

$$\nabla \times (\mathbf{r}\psi_n{}^m) = \nabla \times (\mathbf{r}h_n Y_n{}^m) \approx -(-i)^n \frac{e^{ikr}}{kr}\frac{1}{i}\mathbf{r} \times \nabla Y_n{}^m \tag{160}$$

and

$$\begin{aligned}\nabla \times \nabla \times (\mathbf{r}\psi_n{}^m) &= \nabla \times \nabla \times (\mathbf{r}h_n Y_n{}^m) \\ &\approx -(-i)^n ik \frac{e^{ikr}}{kr}\mathbf{e}_r \times \left(\frac{1}{i}\mathbf{r} \times \nabla Y_n{}^m\right)\end{aligned} \tag{161}$$

Hence in terms of the operator **L** we have

$$\nabla \times (\mathbf{r}\psi_n{}^m) \approx -(-i)^n \frac{e^{ikr}}{kr} \mathbf{L}Y_n{}^m \tag{162}$$

$$\nabla \times \nabla \times (\mathbf{r}\psi_n{}^m) \approx -(-i)^n ik \frac{e^{ikr}}{kr} \mathbf{e}_r \times \mathbf{L}Y_n{}^m \tag{163}$$

Using expressions (162) and (163), we thus see that the basic multipole fields (117) through (120) in the far zone are

$$\mathbf{E}'_{nm} = -(-i)^n i \frac{e^{ikr}}{r} \mathbf{e}_r \times \mathbf{L}Y_n{}^m \tag{164}$$

$$\mathbf{H}'_{nm} = +(-i)^n i \sqrt{\epsilon/\mu} \frac{e^{ikr}}{r} \mathbf{L}Y_n{}^m \tag{165}$$

$$\mathbf{E}''_{nm} = -(-i)^n i \sqrt{\mu/\epsilon} \frac{e^{ikr}}{r} \mathbf{L}Y_n{}^m \tag{166}$$

$$\mathbf{H}''_{nm} = -(-i)^n i \frac{e^{ikr}}{r} \mathbf{e}_r \times \mathbf{L}Y_n{}^m \tag{167}$$

Substituting these multipole fields into expansions (156) and (157), we find that the far-zone electromagnetic field is given by

$$\mathbf{E} = \frac{e^{ikr}}{r} \Big[\sum a_{nm}(-i)^{n+1}\mathbf{e}_r \times \mathbf{L}Y_n{}^m + \sum b_{nm} \sqrt{\mu/\epsilon}\, (-i)^{n+1}\mathbf{L}Y_n{}^m\Big] \tag{168}$$

$$\mathbf{H} = \frac{e^{ikr}}{r} \Big[- \sum a_{nm} \sqrt{\epsilon/\mu}\, (-i)^{n+1}\mathbf{L}Y_n{}^m + \sum b_{nm}(-i)^{n+1}\mathbf{e}_r \times \mathbf{L}Y_n{}^m\Big] \tag{169}$$

Radio-astronomical antennas 5

Observational radio astronomy is concerned with the measurement of the radio waves that are emitted by cosmic radio sources.[1] With the apparently single exception of the monochromatic radiation at $\lambda = 21$ centimeters, i.e., the "hydrogen line" emitted by interstellar hydrogen, cosmic radio waves are rapidly and irregularly varying functions of time, resembling noise. The measurable properties of cosmic radio waves are their direction of arrival, state of polarization, spectrum, and strength. For ground-level observations, radio astronomy is limited essentially to the band ranging approximately from 1 centimeter to 10 meters, because waves of wavelength greater than about 10 meters are unable to penetrate the earth's ionosphere and those of wavelength less than about 1 centimeter are absorbed by the earth's atmospheric gases. However, radio-astronomical observa-

[1] For a popular exposition on radio astronomy, see the delightful and informative monograph by F. G. Smith, "Radio Astronomy," Penguin Books, Inc., Baltimore, 1960. For a comprehensive treatment of the subject, see J. L. Pawsey and R. N. Bracewell, "Radio Astronomy," Oxford University Press, Fair Lawn, N.J., 1955; also I. S. Shklovsky, "Cosmic Radio Waves," Harvard University Press, Cambridge, Mass., 1960. See also F. T. Haddock, Introduction to Radio Astronomy, *Proc. IRE,* **46:** 3 (1958); and R. N. Bracewell, Radio Astronomy Techniques, *Handbuch der Physik,* **LIV,** Springer-Verlag OHG, Berlin. See also J. L. Steinberg and J. Lequeux, "Radio Astronomy," McGraw-Hill Book Company, New York, 1963.

tions have also been made in the band ranging from about 3 millimeters to 1 centimeter.

The instrument that is used to measure cosmic radio waves is the "radio telescope." It consists of three basic components operating in tandem, viz., a receiving antenna, a sensitive receiver, and a recording device. Functionally, the antenna collects the incident radiation and transmits it by means of a wave guide or coaxial line to the input terminals of a receiver; the receiver in turn amplifies and rectifies the input signal; and then the recording device, which is driven by the rectified output of the receiver, presents the data for analysis. The rectified output of the receiver is a measure of the power fed to the receiver by the antenna.

Since the cosmic signals arriving at the input terminals of the receiver are noiselike and similar to the unwanted noise signals which are unavoidably generated by the receiver itself, the receiver must be able to distinguish the desired noise signal from the undesired one. This is a difficult requirement and is met by a "radiometer," which consists of a high-quality receiver and special noise-reducing circuitry. To reduce even further the effects of the receiver noise, radiometers sometimes are supplemented with a low-noise amplifier such as a maser[1] or a parametric amplifier operating in front of the receiver.[2]

The part of the radio telescope that we shall consider in this chapter is the antenna, and our presentation will cover only the radiation theory of such radio-astronomical antennas. The reader interested in the more practical and operational aspects of the subject is referred to the literature.[3]

[1] The first application of a maser (X band) to radio astronomy was made by Giordmaine, Alsop, Mayer, and Townes [J. A. Giordmaine, L. E. Alsop, C. H. Mayer, and C. H. Townes, *Proc. IRE,* **47**: 1062 (1959)]. See also J. V. Jelley and B. F. C. Cooper, An Operational Ruby Maser for Observations at 21 Centimeters with a 60-Foot Radio Telescope, *Rev. Sci. Instr.,* **32**: 166 (1961).

[2] See, for example, F. D. Drake, Radio-astronomy Radiometers and Their Calibration, chap. 12 in G. P. Kuiper and B. M. Middlehurst (eds.), "Telescopes," The University of Chicago Press, Chicago, 1960.

[3] See, for example, J. G. Bolton, Radio Telescopes, chap. 11 in G. P. Kuiper and B. M. Middlehurst (eds.), "Telescopes," The University of Chicago Press, Chicago, 1960.

5.1 Spectral Flux Density

Since an incoming cosmic radio wave is a plane transverse electromagnetic (TEM) wave, its field vectors $\mathbf{E}(\mathbf{r},t)$ and $\mathbf{H}(\mathbf{r},t)$ are perpendicular to each other and to the direction of propagation. Consequently the Poynting vector of the wave, viz., $\mathbf{S}(\mathbf{r},t) = \mathbf{E}(\mathbf{r},t) \times \mathbf{H}(\mathbf{r},t)$, is parallel to the direction of propagation, and its magnitude is given by the quadratic quantity

$$S(\mathbf{r},t) = \sqrt{\epsilon_0/\mu_0}\, \mathbf{E}(\mathbf{r},t) \cdot \mathbf{E}(\mathbf{r},t) \tag{1}$$

It is an observed fact that each component of the field vectors, insofar as its time dependence is concerned, has the character of "noise." That is, at any fixed position $\mathbf{r} = \mathbf{r}_0$ the field vectors are rapidly and irregularly varying functions of time, yet in their gross behavior they are essentially independent of the time and in particular do not vanish at $t = \pm\infty$. They constitute what is known as a stationary random (or stochastic) process.[1] By virtue of this noiselike behavior of the wave, it is the spectral density of the time-average value of $S(\mathbf{r},t)$, and not the instantaneous value of $S(\mathbf{r},t)$ itself, that constitutes a meaningful measure of the strength of the incoming wave.

In order to resolve the incoming signal into its Fourier components, we must introduce the truncated function $\mathbf{E}_T(\mathbf{r}_0,t)$ defined by

$$\begin{aligned} \mathbf{E}_T(\mathbf{r}_0,t) &= \mathbf{E}(\mathbf{r}_0,t) \qquad \text{for } |t| \leq T \\ \mathbf{E}_T(\mathbf{r}_0,t) &= 0 \qquad \text{for } |t| > T \end{aligned} \tag{2}$$

where $2T$ is a long interval of time. Since $\mathbf{E}_T(\mathbf{r}_0,t)$ vanishes at $t = \pm\infty$, its Fourier transform

$$\mathbf{A}_T(\omega) = \frac{1}{2\pi} \int_{-\infty}^{\infty} \mathbf{E}_T(\mathbf{r}_0,t) e^{i\omega t}\, dt \tag{3}$$

[1] For general theory of stochastic (random) processes see, for example, S. O. Rice, Mathematical Analysis of Random Noise, *Bell System Tech. J.*, **23:** 282 (1944); **25:** 46 (1945); S. Chandrasekhar, Stochastic Problems in Physics and Astronomy, *Rev. Mod. Phys.*, **15** (1): 1 (1943).

and its Fourier integral representation

$$\mathbf{E}_T(\mathbf{r}_0,t) = \int_{-\infty}^{\infty} \mathbf{A}_T(\omega)e^{-i\omega t}\,d\omega \tag{4}$$

always exist as long as T is finite. We know from the theory of stochastic processes that the transform $\mathbf{A}_T(\omega)$ increases without bound as $T \to \infty$, whereas the quadratic quantity $|\mathbf{A}_T(\omega)|^2/T$ tends to a definite limit,[1] i.e.,

$$\lim_{T\to\infty} \frac{1}{T} |\mathbf{A}_T(\omega)|^2 = \text{finite limit} \tag{5}$$

This fact suggests that a quadratic quantity such as the time-average Poynting vector be considered. According to Eqs. (1) and (4), the magnitude of Poynting's vector is

$$S_T(\mathbf{r}_0,t) = \sqrt{\frac{\epsilon_0}{\mu_0}} \int_{-\infty}^{\infty} \mathbf{A}_T(\omega')e^{-i\omega' t}\,d\omega' \cdot \int_{-\infty}^{\infty} \mathbf{A}_T(\omega'')e^{-i\omega'' t}\,d\omega'' \tag{6}$$

and its time-average value, defined as

$$\langle S(\mathbf{r}_0,t)\rangle = \lim_{T\to\infty} \frac{1}{2T} \int_{-T}^{T} S_T(\mathbf{r}_0,t)dt \tag{7}$$

is given by

$$\langle S(\mathbf{r}_0,t)\rangle = \sqrt{\frac{\epsilon_0}{\mu_0}} \lim_{T\to\infty} \frac{1}{2T} \int_{-T}^{T} \left[\int_{-\infty}^{\infty} \mathbf{A}_T(\omega')e^{-i\omega' t}\,d\omega' \cdot \int_{-\infty}^{\infty} \mathbf{A}_T(\omega'')e^{-i\omega'' t}\,d\omega''\right] dt \tag{8}$$

Since

$$\lim_{T\to\infty} \frac{1}{2T} \int_{-T}^{T} e^{-i(\omega'+\omega'')t}\,dt = \lim_{T\to\infty} \frac{\pi}{T}\,\delta(\omega' + \omega'') \tag{9}$$

[1] For rigorous mathematical theory see N. Wiener, Generalized Harmonic Analysis, *Acta Math.*, **55:** 117 (1930).

where δ is the Dirac delta function, Eq. (8) reduces to

$$\langle S(\mathbf{r}_0,t)\rangle = \sqrt{\frac{\epsilon_0}{\mu_0}} \lim_{T\to\infty} \frac{\pi}{T} \iint_{-\infty}^{\infty} \mathbf{A}_T(\omega')\cdot\mathbf{A}_T(\omega'')\delta(\omega'+\omega'')d\omega' d\omega'$$

$$= \sqrt{\frac{\epsilon_0}{\mu_0}} \lim_{T\to\infty} \frac{\pi}{T} \int_{-\infty}^{\infty} \mathbf{A}_T(\omega)\cdot\mathbf{A}_T(-\omega)d\omega \qquad (10)$$

Using the relation $\mathbf{A}_T(\omega) = \mathbf{A}_T^*(-\omega)$, which is a consequence of the fact that $\mathbf{E}_T(\mathbf{r}_0,t)$ in Eq. (3) is real, we see that Eq. (10) may be written as the one-sided integral

$$\langle S(\mathbf{r}_0,t)\rangle = \int_0^{\infty} \left[\sqrt{\frac{\epsilon_0}{\mu_0}} \lim_{T\to\infty} \frac{2\pi}{T} \mathbf{A}_T(\omega)\cdot\mathbf{A}_T^*(\omega)\right] d\omega \qquad (11)$$

This expression gives the time-average power density of the incoming wave in watts meter^{-2}; hence the quantity

$$S_\omega(\mathbf{r}_0) \equiv \sqrt{\frac{\epsilon_0}{\mu_0}} \lim_{T\to\infty} \left[\frac{2\pi}{T} \mathbf{A}_T(\omega)\cdot\mathbf{A}_T^*(\omega)\right] \qquad (12)$$

which is known to be finite by virtue of Eq. (5), gives its spectral flux density[1] in watts meter^{-2} (cycles per second)$^{-1}$.

[1] An alternative definition of the spectral flux density $S_\omega(\mathbf{r}_0)$ in terms of the electric field $\mathbf{E}(\mathbf{r}_0,t)$ is based on constructing the autocorrelation function

$$\phi(q) = \sqrt{\frac{\epsilon_0}{\mu_0}} \lim_{T\to\infty} \frac{1}{2T} \int_{-T}^{T} \mathbf{E}(t)\cdot\mathbf{E}(t+q)dt \equiv \sqrt{\frac{\epsilon_0}{\mu_0}} \langle \mathbf{E}(t)\cdot\mathbf{E}(t+q)\rangle$$

and then identifying its Fourier transform with the spectral flux density. That is,

$$S_\omega = \frac{1}{\pi}\int_{-\infty}^{\infty} \phi(q)e^{i\omega q}\,dq = \frac{2}{\pi}\int_0^{\infty} \phi(q)\cos\omega q\,dq$$

If S_ω is to be the spectral density, the integral of S_ω over all frequencies must yield the time-average power. To show that this requirement is met by the above definition, we note that

$$\int_0^{\infty} S_\omega\,d\omega = \frac{2}{\pi}\int_0^{\infty} dq\,\phi(q)\int_0^{\infty}\cos\omega q\,d\omega = 2\int_0^{\infty} dq\,\phi(q)\delta(q) = \phi(0)$$

and recognize that $\phi(0)$ is indeed the time-average power.

From a strictly mathematical viewpoint, to obtain the spectral flux density S_ω one would have to observe the incoming signal for an infinitely long period of time. In practice, obviously, this is neither possible nor desirable. As a practical expediency, the definition is relaxed by choosing T large enough and yet not so large as to iron out significant temporal variations of S_ω. The best that can be done is to observe the incoming signal for successive periods of time equal to the time constant τ of the receiving system. What is observed then is the signal smoothed by successive averaging over finite periods of duration τ. The finiteness of τ produces fluctuations in the record, as does the finiteness of the bandwidth $\Delta\omega$ of the receiving system. The period[1] of the fluctuations is approximately $1/\Delta\omega$; hence in a time interval τ, the incoming signal effectively consists of $n(= \tau\,\Delta\omega)$ independent pulses, whose standard deviation is $1/\sqrt{n} = 1/\sqrt{\tau\,\Delta\omega}$. In view of this we can write

$$\frac{\delta R}{R} = \frac{K}{\sqrt{\tau\,\Delta\omega}} \tag{13}$$

where δR is the standard deviation of the readings R, and K is a dimensionless constant whose value depends on the detailed structure of the receiving system. Thus we see that the finiteness of τ and $\Delta\omega$ produces an uncertainty, or spread, in the readings. Consequently, in order that an incoming signal be detectable, the deflection produced by it must be greater than the deflection δR produced by the inherent fluctuations of the receiving system.

Since S_ω is the power per unit area per unit bandwidth, the power P

[1] To see this, we consider the Fourier integral representation (4) for the frequency band $\omega - \Delta\omega/2$ to $\omega + \Delta\omega/2$ over which $\mathbf{A}_T(\omega)$ is assumed constant. Then

$$\mathbf{E}_T(\mathbf{r}_0,t) = \int_{\omega-\frac{\Delta\omega}{2}}^{\omega+\frac{\Delta\omega}{2}} \mathbf{A}_T(\omega)e^{-i\omega t}\,d\omega = \mathbf{A}_T(\omega)\int_{\omega-\frac{\Delta\omega}{2}}^{\omega+\frac{\Delta\omega}{2}} e^{-i\omega t}\,d\omega$$

$$= 2\mathbf{A}_T(\omega)e^{-i\omega t}\frac{1}{t}\sin\left(t\frac{\Delta\omega}{2}\right)$$

Hence the period of the envelope is $4\pi/\Delta\omega$.

flowing normally through an area A in the frequency range $\omega - \Delta\omega/2$ to $\omega + \Delta\omega/2$ is given by

$$P = AS_\omega \, \Delta\omega \tag{14}$$

5.2 Spectral Intensity, Brightness, Brightness Temperature, Apparent Disk Temperature

In the previous section we defined the spectral flux density S_ω of cosmic radiation. In this section we shall define in terms of S_ω some other useful measures of cosmic radiation.

One of these measures of cosmic radiation is the spectral intensity, defined by

$$dP_\omega = I_\omega(\mathbf{n}')d\sigma \, \mathbf{n}_1 \cdot \mathbf{n}' \, d\Omega(\mathbf{n}') \tag{15}$$

where dP_ω is the radiant power per unit bandwidth flowing through an element of area $d\sigma$ into a solid angle $d\Omega(\mathbf{n}')$, $\mathbf{n}_1$ the unit vector normal to $d\sigma$, and $\mathbf{n}'$ the unit vector along the axis of the solid angle (see Fig. 5.1). The quantity $I_\omega(\mathbf{n}')$ is the spectral intensity of the radiation

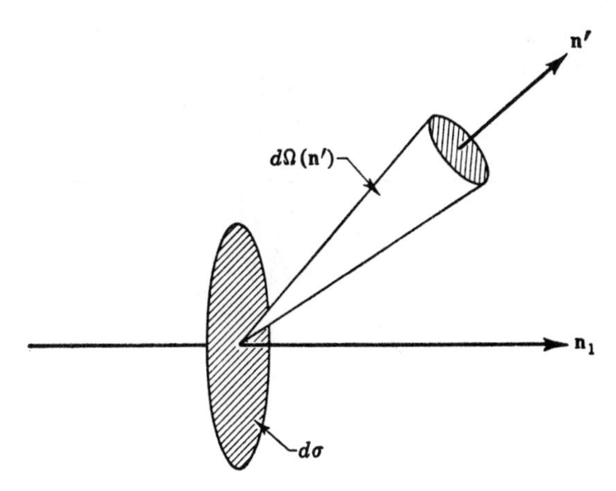

Fig. 5.1 Geometric construction for definition of spectral intensity. Radiation is emitted by source and passes radially outward through solid angle $d\Omega$.

traveling in the direction $\mathbf{n}'$. Another such measure is the spectral brightness, which is defined in the same way as the spectral intensity except that $\mathbf{n}''(= -\mathbf{n}')$ is now the direction from which the radiation is coming (see Fig. 5.2). Accordingly, the power per unit bandwidth falling on the area $d\sigma$ from the solid angle $d\Omega(\mathbf{n}'')$ is given by

$$dP_\omega = b_\omega(\mathbf{n}'')d\sigma\, \mathbf{n}_2 \cdot \mathbf{n}''\, d\Omega(\mathbf{n}'') \tag{16}$$

where $b_\omega(\mathbf{n}'')$ is the spectral brightness of the incoming radiation and $\mathbf{n}_2 = -\mathbf{n}_1$. Comparing expressions (15) and (16), we get the relation

$$b_\omega(-\mathbf{n}')d\Omega(-\mathbf{n}') = I_\omega(\mathbf{n}')d\Omega(\mathbf{n}') \tag{17}$$

which places in evidence the fact that brightness refers to radiation traveling toward $d\sigma$ and intensity refers to radiation traveling away from $d\sigma$.

The quantity $dP_\omega/(d\sigma\, \mathbf{n}_1 \cdot \mathbf{n}')$ is the power per unit bandwidth per unit area normal to the direction of travel of the radiation and hence it is equal to dS_ω. Thus we see that the spectral intensity of the

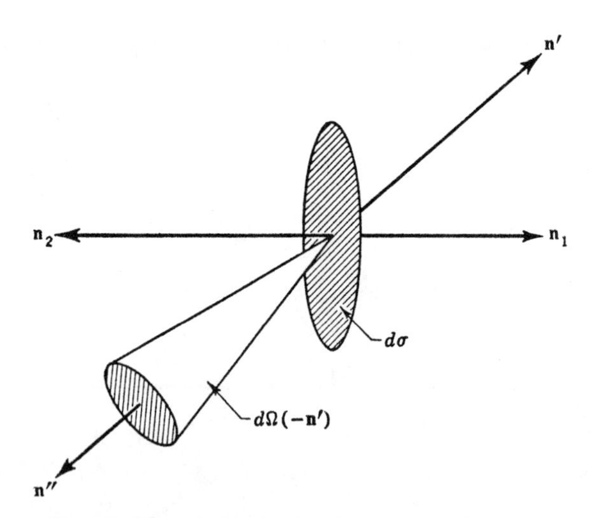

Fig. 5.2 Geometric construction for definition of brightness. Radiation is emitted by distributed source in sky and passes radially inward through solid angle $d\Omega$.

emitted radiation is the spectral flux density per unit solid angle, i.e.,

$$I_\omega = \frac{dS_\omega}{d\Omega} \tag{18}$$

Similarly the spectral brightness of the received radiation is the spectral flux density per unit solid angle, i.e.,

$$b_\omega = \frac{dS_\omega}{d\Omega} \tag{19}$$

If the source of radiation is distributed over the sky, then a convenient measure of the amount of radiation that falls on a receiving antenna from a given direction is the spectral brightness in that direction. As shown in Figs. 5.1 and 5.2, it is most convenient to choose the origin of coordinates at the source for I_ω, and at the receiver for b_ω. The spectral brightness b_ω, like the spectral intensity I_ω, is a function of θ, ϕ but not of r.

The units of I_ω and b_ω are the same since they are defined in the same way, except that in the former the radiation is traveling outward from the vertex of the solid angle where the source is located, and in the latter it is traveling in toward the vertex where the receiver is located. Specifically, the units of I_ω and b_ω are watts meter^{-2} (cycles per second)$^{-1}$ steradian^{-1}.

It is sometimes convenient to specify the radiation in terms of the temperature that a blackbody would require in order to produce the measured spectral brightness. According to Planck's law for blackbody radiation in free space, the spectral brightness B_ω of a blackbody at temperature T (in degrees Kelvin) is given by

$$B_\omega = \frac{2hc}{\lambda^3} \frac{1}{\exp(hc/k\lambda T) - 1} \tag{20}$$

where

h = Planck's constant
k = Boltzmann's constant
c = velocity of light
λ = wavelength

But in radio-astronomical applications $hc \ll k\lambda T$ and hence expression (20) may be replaced by the Rayleigh-Jeans approximation to Planck's law, viz.,

$$B_\omega = \frac{2kT}{\lambda^2} \tag{21}$$

The spectral brightness temperature $T_{\omega b}$ of the radiation coming toward the receiving antenna along a direction θ, ϕ is obtained by equating expression (21) to b_ω. Thus the spectral brightness temperature $T_{\omega b}$ is related to the spectral brightness by

$$b_\omega = \frac{2k}{\lambda^2} T_{\omega b} = 2.77 \times 10^{-23} T_{\omega b}/\lambda^2 \tag{22}$$

Like b_ω, the quantity $T_{\omega b}$ is a function of θ, ϕ only. In case the source subtends a solid angle Ω_0 at the receiver, it follows from Eqs. (19) and (22) that

$$S_\omega = \int_{\Omega_0} b_\omega \, d\Omega = \frac{2k}{\lambda^2} \int_{\Omega_0} T_{\omega b} \, d\Omega \tag{23}$$

Noting that the "apparent disk temperature" $T_{\omega d}$ is defined by

$$S_\omega = \frac{2k}{\lambda^2} \int_{\Omega_0} T_{\omega d} \, d\Omega \tag{24}$$

we see from Eq. (23) that $T_{\omega d}$ is related to $T_{\omega b}$ by

$$T_{\omega d} = \frac{1}{\Omega_0} \int_{\Omega_0} T_{\omega b} \, d\Omega \tag{25}$$

and in this sense constitutes a measure of the average value of the spectral brightness temperature.

5.3 Poincaré Sphere, Stokes Parameters

By its very nature a monochromatic electromagnetic wave must be elliptically polarized, i.e., the end point of its electric vector at each point of space must trace out periodically an ellipse or one of its special forms, viz., a circle or straight line. On the other hand, a polychromatic electromagnetic wave can be in any state of polarization, ranging from the elliptically polarized state to the unpolarized state, wherein the end point of the electric vector moves quite irregularly. Cosmic radio waves are generally in neither of these two extreme states, but rather in an intermediate state containing both elliptically polarized and unpolarized parts. A wave in such an intermediate state is said to be "partially polarized" and is describable by four parameters introduced by Sir George Gabriel Stokes in 1852 in connection with his investigations of partially polarized light.[1] In this section we define these Stokes parameters and show that they serve as a complete measure of the state of polarization.

As an exemplar, we consider the case of a plane monochromatic TEM wave. The electric vector $\mathbf{E}(\mathbf{r},t)$ of such a wave traveling in the direction of the unit vector $\mathbf{n}$ has the form

$$\mathbf{E}(\mathbf{r},t) = \mathrm{Re}\,\{\mathbf{E}_0 e^{i(\mathbf{k}\cdot\mathbf{r}-\omega t)}\} \tag{26}$$

where $\mathbf{k}(=\mathbf{n}2\pi/\lambda)$ is the wave vector and $\mathbf{E}_0$ the complex vector amplitude. Because the wave is plane and TEM, vector $\mathbf{E}_0$ is a constant and lies in a plane perpendicular to $\mathbf{n}$, that is, $\mathbf{n}\cdot\mathbf{E}_0 = 0$. Since the polarization of the wave is governed by $\mathbf{E}_0$, and since $\mathbf{E}_0$ is a constant, the state of polarization is the same everywhere. This constancy of polarization is peculiar to plane, homogeneous waves, the polarization of more general types of electromagnetic fields possibly being different at different points of space. For example, if the field were a wave generated by a source of finite spatial extent, the polarization would vary with radial distance from the source as well as with polar and azimuthal angles.

Without sacrificing generality, we choose a cartesian coordinate system x, y, z such that the z axis is parallel to $\mathbf{n}$. With respect to this system $\mathbf{E}_0$ can be written as

$$\mathbf{E}_0 = \mathbf{e}_x a_x e^{-i\delta_x} + \mathbf{e}_y a_y e^{-i\delta_y} \tag{27}$$

[1] G. G. Stokes, On the Composition and Resolution of Streams of Polarized Light from Different Sources, *Trans. Cambridge Phil. Soc.*, **9:** 399 (1852). Reprinted in his "Mathematical and Physical Papers," vol. III, pp. 233–258, Cambridge University Press, London.

where $\mathbf{e}_x$, $\mathbf{e}_y$ are unit vectors along the x and y axes respectively, and where the amplitudes a_x, a_y as well as the phases δ_x, δ_y are real constants. Thus from Eqs. (26) and (27) it follows that the cartesian components of $\mathbf{E}(z,t)$ are given by the real expressions

$$E_x = a_x \cos(\phi + \delta_x) \qquad E_y = a_y \cos(\phi + \delta_y) \qquad E_z = 0 \tag{28}$$

where for brevity the shorthand $\phi = \omega t - kz$ has been used. Eliminating ϕ from these expressions, we get

$$\left(\frac{E_x}{a_x}\right)^2 + \left(\frac{E_y}{a_y}\right)^2 - 2\frac{E_x}{a_x}\frac{E_y}{a_y}\cos\delta = \sin^2\delta \tag{29}$$

where

$$\delta = \delta_y - \delta_x \tag{30}$$

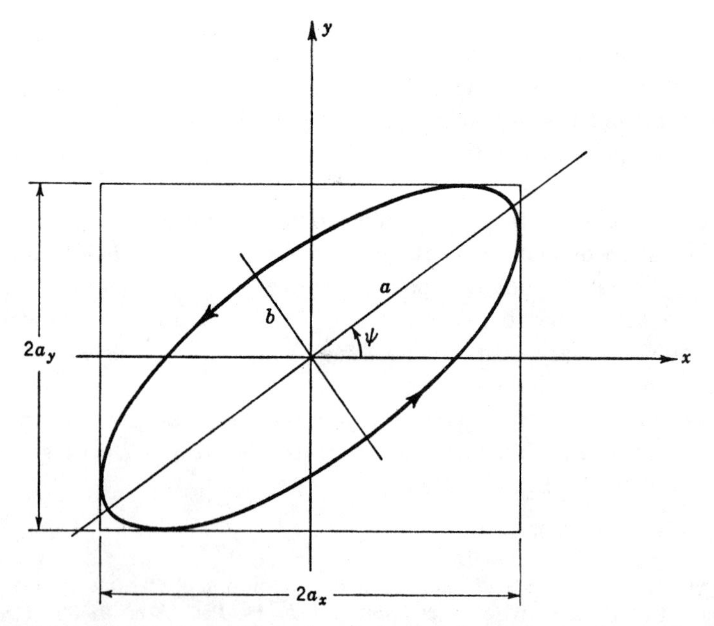

Fig. 5.3 ***Polarization ellipse for right-handed polarized wave having orientation angle ψ.***

Taking E_x and E_y as coordinate axes, we see that Eq. (29) represents an ellipse whose center is at the origin $E_x = E_y = 0$. Geometrically this means that at each point of space the vector $\mathbf{E}$ rotates in a plane perpendicular to $\mathbf{n}$ and in so doing traces out an ellipse. As is evident from expressions (28), the rotation of $\mathbf{E}$ and the direction of propagation $\mathbf{n}$ form either a right-handed screw or a left-handed screw, depending on whether $\sin\delta < 0$ or $\sin\delta > 0$ respectively. Accordingly, in conformity to standard radio terminology, the polarization of a wave receding from the observer is called right-handed if the electric vector appears to be rotating clockwise and left-handed if it appears to be rotating counterclockwise. See Fig. 5.3.

To determine the polarization ellipse of a monochromatic wave, a set of three independent quantities is needed. One such set obviously consists of the amplitudes a_x, a_y and the phase difference δ. Another set is made up of the semimajor and semiminor axes of the ellipse, denoted by a and b respectively, and the orientation angle ψ between the major axis of the ellipse and the x axis of the coordinate system. These two sets are related such that a, b, ψ can be found from a_x, a_y, δ and vice versa. The well-known connection relations are[1]

$$a^2 + b^2 = a_x^2 + a_y^2 \tag{31}$$

$$\tan 2\psi = \frac{2a_x a_y}{a_x^2 - a_y^2} \cos\delta \qquad (0 \leq \psi < \pi) \tag{32}$$

Moreover, we have

$$\sin 2\chi = \frac{2a_x a_y}{a_x^2 + a_y^2} \sin\delta \tag{33}$$

where χ is an auxiliary angle defined by

$$\tan\chi = \pm\frac{b}{a} \qquad (-\pi/4 \leq \chi \leq \pi/4) \tag{34}$$

The numerical value of $\tan\chi$ yields the reciprocal of the axial ratio a/b of the ellipse, and the sign of χ differentiates the two senses of polariza-

[1] See, for example, M. Born and E. Wolf, "Principles of Optics," pp. 24–31, Pergamon Press, New York, 1959.

tion, e.g., for left-handed polarization $0 < \chi \leq \pi/4$ and for right-handed polarization $-\pi/4 \leq \chi < 0$.

The Stokes parameters for the monochromatic plane TEM wave (28) are the four quantities

$$s_0 = a_x^2 + a_y^2 \qquad s_1 = a_x^2 - a_y^2 \qquad s_2 = 2a_xa_y \cos \delta \qquad s_3 = 2a_xa_y \sin \delta \tag{35}$$

But since the quantities are related by the identity

$$s_0^2 = s_1^2 + s_2^2 + s_3^2 \tag{36}$$

only three of the four parameters are independent. Alternatively, the Stokes parameters can be written in terms of the orientation angle ψ and the ellipticity angle χ as follows:

$$s_1 = s_0 \cos 2\chi \cos 2\psi \qquad s_2 = s_0 \cos 2\chi \sin 2\psi \qquad s_3 = s_0 \sin 2\chi \tag{37}$$

where s_0 is proportional to the intensity of the wave. From these expressions we see that if s_1, s_2, s_3 are interpreted as the cartesian coordinates of a point on a sphere of radius s_0, known as the Poincaré sphere,[1] the longitude and latitude of the point are 2ψ and 2χ respectively (see Fig. 5.4). Thus, there is a one-to-one correspondence between the points on the sphere and the states of polarization of the wave. In order that the wave be linearly polarized, the phase difference δ must be zero or an integral multiple of π, and consequently, according to Eq. (33), χ must be zero. Thus we see that the points on the equator of the Poincaré sphere correspond to linearly polarized waves. In order that the wave be circularly polarized, the amplitudes a_x and a_y must be equal and the phase difference δ must be either $\pi/2$ or $-\pi/2$, depending on whether the sense of polarization is left-handed or right-handed respectively. Hence, from Eq. (33) it follows that for a left-handed circularly polarized wave $2\chi = \pi/2$ and for a right-handed circularly polarized wave $2\chi = -\pi/2$; that is, the north and south poles of the Poincaré sphere correspond respectively to left-handed and right-handed circular polarization. The other points on the Poincaré sphere

[1] H. Poincaré, "Théorie mathématique de la lumière," vol. 2, chap. 12, Paris, 1892.

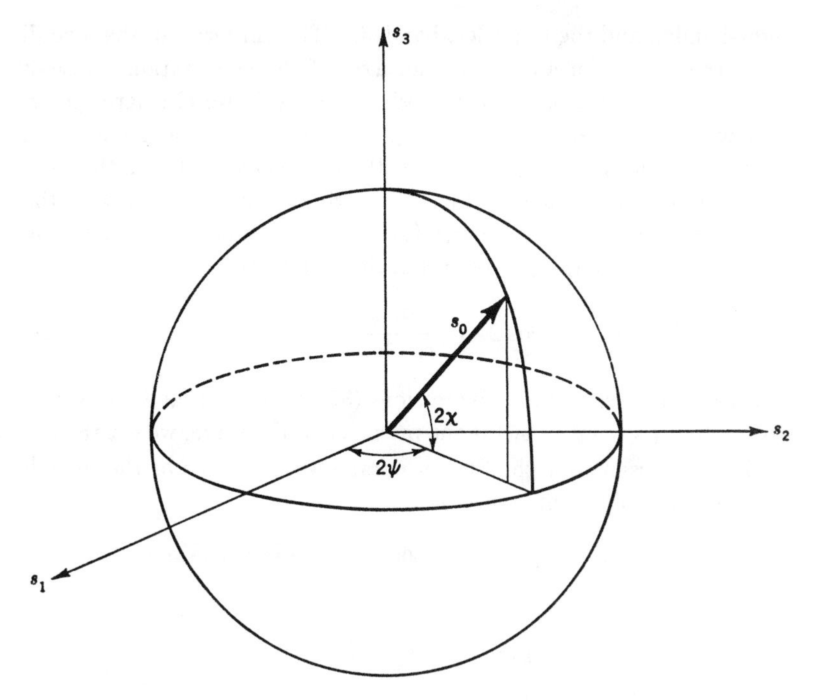

***Fig. 5.4** Poincaré sphere is a sphere of radius s_0. A point on sphere has latitude 2χ and longitude 2ψ.*

represent elliptic polarization, right-handed in the southern hemisphere and left-handed in the northern hemisphere.

Since $\mathbf{E}(z,t)$ has only two components E_x, E_y it can be represented, for any fixed value of $z(= z_0)$, as a vector in the complex plane whose real and imaginary axes are E_x and E_y respectively. That is, as a function of t, to each value of $\mathbf{E}(z_0,t)$ there corresponds a point $E_x + iE_y$ in the Argand diagram.[1] With the aid of this representation an elliptically polarized wave may be decomposed into a right-handed and a left-handed circularly polarized wave. We note that in the complex plane circularly polarized waves of opposite senses are given by the complex vectors $\rho_1 \exp(i\omega t)$ and $\rho_2 \exp(-i\omega t + i\gamma)$, the former being

[1] See, for example, K. C. Westfold, New Analysis of the Polarization of Radiation and the Faraday Effect in Terms of Complex Vectors, *J. Opt. Soc. Am.*, **49:** 717 (1959).

right-handed and the latter left-handed. Thus in terms of the moduli ρ_1, ρ_2 the semimajor and semiminor axes of the polarization ellipse of the wave consisting of the superposition of these two circularly polarized waves are $\rho_2 + \rho_1$ and $\rho_2 - \rho_1$. The orientation angle ψ of the ellipse is given by $2\psi = \gamma$, where γ is the phase angle between the complex vectors at $t = 0$ (see Fig. 5.5). Since the axial ratio of the polarization ellipse is $(\rho_2 - \rho_1)/(\rho_2 + \rho_1)$, the angle χ is given by $\tan \chi = (\rho_2 - \rho_1)/(\rho_2 + \rho_1)$. From this it follows that

$$\sin 2\chi = \frac{\rho_2^2 - \rho_1^2}{\rho_2^2 + \rho_1^2} \qquad \cos 2\chi = \frac{2\rho_2\rho_1}{\rho_2^2 + \rho_1^2} \tag{38}$$

Substituting Eqs. (38) into expressions (37), recalling that $2\psi = \gamma$, and noting that $s_0(= \rho_2^2 + \rho_1^2)$ is the intensity of the wave, we get the following expressions for the Stokes parameters in terms of the moduli ρ_2, ρ_1 and the phase difference γ:

$$s_0 = 2(\rho_2^2 + \rho_1^2) \qquad s_1 = 4\rho_2\rho_1 \cos \gamma \qquad s_2 = 4\rho_2\rho_1 \sin \gamma$$

$$s_3 = 2(\rho_2^2 - \rho_1^2) \tag{39}$$

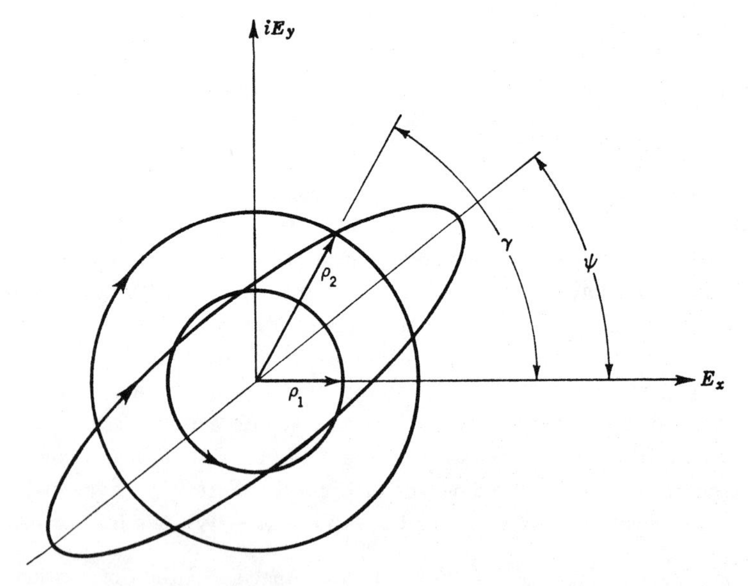

Fig. 5.5 Complex plane. Splitting of elliptical polarization into two oppositely polarized circular components.

The state of polarization can be measured in a number of different ways. For example, as is suggested by expressions (35), the state of polarization can be measured by using two linearly polarized receiving antennas in such a way that one yields a_x, the other yields a_y, and the phase difference between their responses yields δ. Alternatively, from expressions (39) it is seen that the state of polarization also can be measured by using two circularly polarized antennas of opposite senses, one of the antennas yielding ρ_2, the other yielding ρ_1, and the phase difference of their responses yielding γ. The accuracy of these methods of measurement depends largely on how purely linear is the linearly polarized antenna and how purely circular is the circularly polarized antenna. The techniques of measuring the polarization of monochromatic waves are well known and will not be discussed here.[1]

Using the monochromatic wave (26) as a prototype, we now examine the case of a plane polychromatic TEM wave, which by virtue of its polychromatic character can be elliptically polarized, or unpolarized, or partially polarized. We assume that the frequency spectrum of the wave is confined to a relatively narrow band of width $\Delta\omega$ so that the electric vector of the wave, in analogy with expression (26), may have the simple analytic representation

$$\mathbf{E}(z,t) = \operatorname{Re}\,\{\mathbf{E}_0(t)e^{i(z\omega/c-\omega t)}\} \tag{40}$$

where ω now denotes some average value of the frequency. Because the bandwidth is narrow, $\mathbf{E}_0(t)$ may change by only a relatively small amount in the time interval $1/\Delta\omega$ and in this sense is a slowly varying function of time. If the bandwidth were unrestricted, the moot question of representing a broadband signal in analytical form would arise and the problem would have to be reformulated.[2] In practice, how-

[1] See, for example, H. G. Booker, V. H. Rumsey, G. A. Deschamps, M. L. Kales, and J. I. Bohnert, Techniques for Handling Elliptically Polarized Waves with Special Reference to Antennas, *Proc. IRE*, **39:** 533 (1951); D. D. King, "Measurements at Centimeter Wavelength," pp. 298–309, D. Van Nostrand Company, Inc., Princeton, N.J., 1950; J. D. Kraus, "Antennas," pp. 479–484, McGraw-Hill Book Company, New York, 1950.

[2] See A. D. Jacobson, Theory of Noise-like Electromagnetic Fields of Arbitrary Spectral Width, *Caltech Antenna Lab. Report*, No. 32, June, 1964. Also, Robert M. Lerner, Representation of Signals, chap. 10 in E. J. Baghdady (ed.), "Lectures on Communication System Theory," McGraw-Hill Book Company, 1961.

ever, this difficulty is compulsorily bypassed inasmuch as the instruments used in measuring polarization are inherently narrowband devices.

Writing $\mathbf{E}_0(t)$ in the form

$$\mathbf{E}_0(t) = \mathbf{e}_x a_x(t) e^{-i\delta_x(t)} + \mathbf{e}_y a_y(t) e^{-i\delta_y(t)} \tag{41}$$

where the amplitudes $a_x(t)$, $a_y(t)$ and the phases $\delta_x(t)$, $\delta_y(t)$ are slowly varying functions of time, we see from Eq. (40) that the cartesian components of $\mathbf{E}(z,t)$ are given by

$$\begin{aligned} E_x &= a_x(t) \cos [\phi + \delta_x(t)] \\ E_y &= a_y(t) \cos [\phi + \delta_x(t) + \delta(t)] \\ E_z &= 0 \end{aligned} \tag{42}$$

where $\phi = \omega t - z\omega/c$, $\delta(t) = \delta_y(t) - \delta_x(t)$. Although the amplitudes and phases are irregularly varying functions of time, certain correlations may exist among them. It is these correlations that determine the Stokes parameters and consequently the polarization of the wave. By definition, the Stokes parameters of the polychromatic wave (42) are the time-averaged quantities

$$\begin{aligned} s_0 &= \langle a_x^2(t) \rangle + \langle a_y^2(t) \rangle \qquad & s_1 &= \langle a_x^2(t) \rangle - \langle a_y^2(t) \rangle \\ s_2 &= 2\langle a_x(t) a_y(t) \cos \delta(t) \rangle \qquad & s_3 &= 2\langle a_x(t) a_y(t) \sin \delta(t) \rangle \end{aligned} \tag{43}$$

which are generalizations of the monochromatic Stokes parameters (35). It can be shown[1] that the polychromatic Stokes parameters satisfy the relation

$$s_0^2 \geq s_1^2 + s_2^2 + s_3^2 \tag{44}$$

where the equality sign holds only when the polychromatic wave is elliptically polarized.

The polychromatic wave (42) is elliptically polarized when the ratio q of the amplitudes ($q = a_y/a_x$) and the phase differences δ are absolute

[1] See, for example, S. Chandrasekhar, "Radiative Transfer," pp. 24–34, Dover Publications, Inc., New York, 1960.

constants. That is, when q and δ are time-independent, the electric vector of the wave traces out an ellipse whose size continually varies at a rate controlled by the bandwidth $\Delta\omega$ but whose shape, orientation, and sense of polarization do not change. To demonstrate this, we note that for an elliptically polarized wave the Stokes parameters (43) become

$$s_0 = (1 + q^2)\langle a_x^2(t)\rangle \qquad s_1 = (1 - q^2)\langle a_x^2(t)\rangle$$
$$s_2 = 2q\langle a_x^2(t)\rangle \cos \delta \qquad s_3 = 2q\langle a_x^2(t)\rangle \sin \delta \tag{45}$$

Since these parameters satisfy the identity $s_0^2 = s_1^2 + s_2^2 + s_3^2$, only three of them are independent. In analogy with Eqs. (37) we can write the Stokes parameters of an elliptically polarized polychromatic wave in the form

$$s_1 = s_0 \cos 2\chi \cos 2\psi \qquad s_2 = s_0 \cos 2\chi \sin 2\psi \qquad s_3 = s_0 \sin 2\chi \tag{46}$$

Consequently the orientation angle ψ of the polarization ellipse is given by

$$\tan 2\psi = \frac{s_2}{s_1} = \frac{2q}{1 - q^2} \cos \delta \tag{47}$$

and its ellipticity angle χ by

$$\sin 2\chi = \frac{s_3}{s_0} = \frac{2q}{1 + q^2} \sin \delta \tag{48}$$

Since q and δ are time-independent, it is clear from Eqs. (47) and (48) that ψ and χ are time-independent, in confirmation of the fact that the shape, orientation, and sense of polarization do not change.

We return to the polychromatic wave (42) and now assume that the phase of E_y is shifted with respect to the phase of E_x by an arbitrary constant amount ϵ. The cartesian components of such a polychromatic wave are given by

$$E_x = a_x(t) \cos [\phi + \delta_x(t)]$$
$$E_y = a_y(t) \cos [\phi + \delta_x(t) + \delta(t) + \epsilon] \tag{49}$$
$$E_z = 0$$

As is clear from Fig. 5.6, the component of the electric field along the x' axis making an angle θ with the x axis is $E_{x'}(\theta,\epsilon) = E_x \cos\theta + E_y \sin\theta$ and its square is $E_{x'}^2(\theta,\epsilon) = E_x^2 \cos^2\theta + E_y^2 \sin^2\theta + 2E_xE_y \cos\theta \sin\theta$. Substituting expressions (49) into this quadratic form, we find that the instantaneous value of $E_{x'}^2(\theta,\epsilon)$ is

$$E_{x'}^2(\theta,\epsilon) = a_x^2(t)\cos^2\tau\cos^2\theta + a_y^2(t)\cos^2[\tau + \delta(t) + \epsilon]\sin^2\theta + 2a_x(t)a_y(t)\cos\tau\cos[\tau + \delta(t) + \epsilon]\cos\theta\sin\theta$$

where $\tau = \phi + \delta_x(t)$. Recalling that $a_x(t)$, $a_y(t)$, $\delta(t)$ are slowly varying functions of time and that

$$\cos\tau = \cos[\phi + \delta_x(t)] = \cos[\omega t - z\omega/c + \delta_x(t)]$$

is a rapidly varying function of time, we find that the mean value of $2E_{x'}^2(\theta,\epsilon)$, which we denote by $I(\theta,\epsilon)$, has the following representation:

$$I(\theta,\epsilon) = 2\langle E_{x'}^2(\theta,\epsilon)\rangle = \langle a_x^2(t)\rangle\cos^2\theta + \langle a_y^2(t)\rangle\sin^2\theta + [\langle a_x(t)a_y(t)\cos\delta(t)\rangle\cos\epsilon - \langle a_x(t)a_y(t)\sin\delta(t)\rangle\sin\epsilon]\sin 2\theta \quad (50)$$

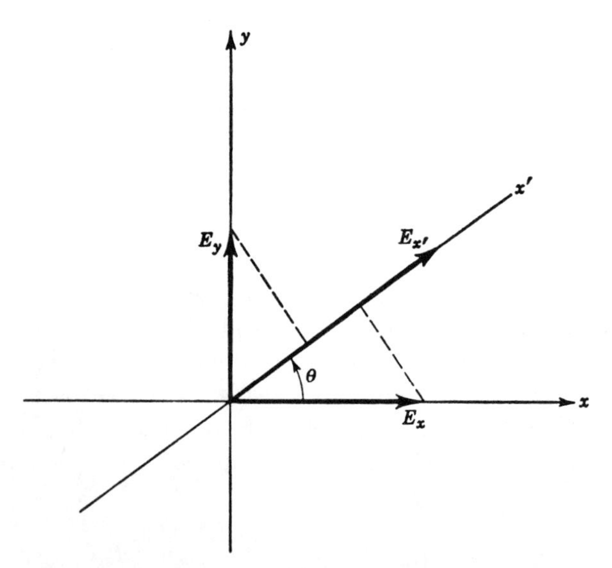

Fig. 5.6 Linearly polarized antenna that picks up the component E_x of electric field along x' axis. Its response is proportional to the mean-square value of $E_{x'}$.

With the aid of definitions (43) this representation leads directly to the relation

$$I(\theta,\epsilon) = \tfrac{1}{2}[s_0 + s_1 \cos 2\theta + (s_2 \cos \epsilon - s_3 \sin \epsilon) \sin 2\theta] \tag{51}$$

which shows that $I(\theta,\epsilon)$ is linearly related to the Stokes parameters. It is evident from relation (51) that the Stokes parameters can be determined by measuring $I(\theta,\epsilon)$ for various values of θ and ϵ.

If $I(\theta,\epsilon)$ happens to be independent of θ and ϵ, the wave is said to be "unpolarized." In other words, an unpolarized wave is one that satisfies

$$I(\theta,\epsilon) = \tfrac{1}{2}s_0 \tag{52}$$

independently of θ and ϵ, or, equivalently, the necessary and sufficient condition that the wave be unpolarized is

$$s_1 = s_2 = s_3 = 0 \tag{53}$$

If the polychromatic wave consists of a superposition of several physically independent waves, the intensity of the resulting wave is the sum of the intensities of the independent waves. That is, if $I^{(n)}$ denotes the intensity of the nth independent wave, the intensity I of the composite wave is given by

$$I = \sum_n I^{(n)} \tag{54}$$

Moreover, since each of the independent waves satisfies relation (51), we have for the nth independent wave

$$I^{(n)}(\theta,\epsilon) = \tfrac{1}{2}[s_0^{(n)} + s_1^{(n)} \cos 2\theta + (s_2^{(n)} \cos \epsilon - s_3^{(n)} \sin \epsilon) \sin 2\theta] \tag{55}$$

where $s_0^{(n)}$, $s_1^{(n)}$, $s_2^{(n)}$, $s_3^{(n)}$ are the corresponding Stokes parameters. Hence, from Eqs. (54) and (55) we get the expression

$$I(\theta,\epsilon) = \tfrac{1}{2}[\Sigma s_0^{(n)} + \Sigma s_1^{(n)} \cos 2\theta + (\Sigma s_2^{(n)} \cos \epsilon - \Sigma s_3^{(n)} \sin \epsilon) \sin 2\theta] \tag{56}$$

which, when compared with expression (51), shows that each of the Stokes parameters of the composite wave is the sum of the respective Stokes parameters of the independent waves. That is, the Stokes parameters are additive in the sense that

$$s_0 = \Sigma s_0^{(n)} \qquad s_1 = \Sigma s_1^{(n)} \qquad s_2 = \Sigma s_2^{(n)} \qquad s_3 = \Sigma s_3^{(n)} \tag{57}$$

where s_0, s_1, s_2, s_3 are the Stokes parameters of the composite wave and $s_0^{(n)}$, $s_1^{(n)}$, $s_2^{(n)}$, $s_3^{(n)}$ ($n = 1, 2, 3, \ldots$) are the Stokes parameters of the independent waves into which the composite wave can be decomposed.

With the aid of this additivity of the Stokes parameters we can show that a polychromatic wave is decomposable uniquely into an unpolarized part and an elliptically polarized part, the two parts being mutually independent. To do this, we denote the Stokes parameters of the composite wave by (s_0,s_1,s_2,s_3), those of the unpolarized part by $(s_0^{(1)},0,0,0)$, and those of the polarized part by $(s_0^{(2)},s_1^{(2)},s_2^{(2)},s_3^{(2)})$. Then by the additivity relation (57) we have

$$s_0 = s_0^{(1)} + s_0^{(2)} \qquad s_1 = s_1^{(2)} \qquad s_2 = s_2^{(2)} \qquad s_3 = s_3^{(2)} \tag{58}$$

The degree of polarization m is defined as the ratio of the intensity of the polarized part to the intensity of the composite wave, i.e., by definition

$$m = \frac{s_0^{(2)}}{s_0} \tag{59}$$

From relation (44) we know that the Stokes parameters of the polarized part are connected by the relation

$$[s_0^{(2)}]^2 = [s_1^{(2)}]^2 + [s_2^{(2)}]^2 + [s_3^{(2)}]^2 \tag{60}$$

which, with the aid of the last three equalities of Eq. (58), can be written as

$$[s_0^{(2)}]^2 = s_1^2 + s_2^2 + s_3^2 \tag{61}$$

It follows from definition (59) and relation (61) that in terms of the Stokes parameters of the composite wave the degree of polarization is given by

$$m = \frac{(s_1^2 + s_2^2 + s_3^2)^{1/2}}{s_0} \tag{62}$$

Furthermore, the orientation of the polarization ellipse is given by

$$\tan 2\psi = \frac{s_2^{(2)}}{s_1^{(2)}} = \frac{s_2}{s_1} \tag{63}$$

and its ellipticity by

$$\sin 2\chi = \frac{s_3^{(2)}}{s_0^{(2)}} = \frac{s_3}{(s_1^2 + s_2^2 + s_3^2)^{1/2}} \tag{64}$$

where use has been made of Eqs. (58) and (61). Thus we see that when the Stokes parameters s_0, s_1, s_2, s_3 of a partially polarized wave are known we can calculate the degree of polarization from Eq. (62), and the properties of the polarization ellipse of the polarized part of the wave from Eqs. (63) and (64). Since χ is restricted to the interval $-\pi/4 \leq \chi \leq \pi/4$, Eq. (64) unambiguously yields a single value for χ. However, Eq. (63) can be satisfied by two values of ψ differing by $\pi/2$, the restriction that ψ lie in the interval $0 \leq \psi \leq \pi$ not being sufficient to fix ψ unambiguously. But from the first two of Eqs. (46) we see that ψ must be chosen such that s_1 and s_2 have the proper signs. Consequently, ψ is determined by Eq. (63) and by the requirement that the appropriate part of the Poincaré sphere be used.

Another way of decomposing a polychromatic wave is to express it as the superposition of two oppositely polarized independent waves. Two waves are said to be "oppositely polarized" if the orientation and ellipticity angles ψ_1, χ_1 of one of the waves are related as follows to the orientation and ellipticity angles ψ_2, χ_2 of the other wave:

$$\chi_1 = -\chi_2 \qquad \psi_2 = \psi_1 + \frac{\pi}{2} \tag{65}$$

This means that the major axes of the polarization ellipses of oppositely polarized waves are perpendicular to each other, that the axial ratios of the ellipses are equal, and that the senses of polarization are opposite.

Let (s_0,s_1,s_2,s_3) denote the Stokes parameters of the polychromatic waves, and let $(s_0^{(1)},s_1^{(1)},s_2^{(1)},s_3^{(1)})$ and $(s_0^{(2)},s_1^{(2)},s_2^{(2)},s_3^{(2)})$ denote the Stokes parameters of the two independent and oppositely polarized waves. By the additivity of the Stokes parameters we have

$$s_0 = s_0^{(1)} + s_0^{(2)} \tag{66}$$

This relation is satisfied if we choose

$$\begin{aligned} s_0^{(1)} &= \tfrac{1}{2}s_0 - \alpha \\ s_0^{(2)} &= \tfrac{1}{2}s_0 + \alpha \end{aligned} \tag{67}$$

where α is an unknown quantity. Then the Stokes parameters of the oppositely polarized waves are

$$\begin{aligned} &(\tfrac{1}{2}s_0 - \alpha) \qquad (\tfrac{1}{2}s_0 - \alpha)\cos 2\chi_1 \cos 2\psi_1 \\ &(\tfrac{1}{2}s_0 - \alpha)\cos 2\chi_1 \sin 2\psi_1 \qquad (\tfrac{1}{2}s_0 - \alpha)\sin 2\chi_1 \end{aligned} \tag{68}$$

and

$$\begin{aligned} &(\tfrac{1}{2}s_0 + \alpha) \qquad (\tfrac{1}{2}s_0 + \alpha)\cos 2\chi_2 \cos 2\psi_2 \\ &(\tfrac{1}{2}s_0 + \alpha)\cos 2\chi_2 \sin 2\psi_2 \qquad (\tfrac{1}{2}s_0 + \alpha)\sin 2\chi_2 \end{aligned} \tag{69}$$

Since these waves are oppositely polarized, we have

$$\begin{aligned} &\cos 2\chi_1 = \cos 2\chi_2 \qquad \sin 2\chi_1 = -\sin 2\chi_2 \\ &\cos 2\psi_1 = -\cos 2\psi_2 \qquad \sin 2\psi_1 = -\sin 2\psi_2 \end{aligned} \tag{70}$$

In view of these relations we see that if the additivity theorem is applied to the Stokes parameters of the original wave and to the Stokes parameters (68) and (69) of the two oppositely polarized waves, the following relations are obtained:

$$\begin{aligned} -2\alpha \cos 2\chi_1 \cos 2\psi_1 &= s_1 \\ -2\alpha \cos 2\chi_1 \sin 2\psi_1 &= s_2 \\ -2\alpha \sin 2\chi_1 &= s_3 \end{aligned} \tag{71}$$

Squaring and adding Eqs. (71), we obtain

$$4\alpha^2 = s_1{}^2 + s_2{}^2 + s_3{}^2 \tag{72}$$

From Eqs. (67) and (72) it follows that the intensities of the two oppositely polarized waves are

$$\begin{aligned} s_0{}^{(1)} &= \tfrac{1}{2}s_0 - \tfrac{1}{2}\sqrt{s_1{}^2 + s_2{}^2 + s_3{}^2} \\ s_0{}^{(2)} &= \tfrac{1}{2}s_0 + \tfrac{1}{2}\sqrt{s_1{}^2 + s_2{}^2 + s_3{}^2} \end{aligned} \tag{73}$$

From Eqs. (71) we see that ψ_1 and χ_1 are given by

$$\tan 2\psi_1 = \frac{s_2}{s_1} \qquad \sin 2\chi_1 = \frac{-s_3}{\sqrt{s_1{}^2 + s_2{}^2 + s_3{}^2}} \tag{74}$$

Thus we see that a polychromatic wave whose Stokes parameters are (s_0,s_1,s_2,s_3) can be decomposed into two polarized waves having the intensities $(\tfrac{1}{2})s_0 \pm (\tfrac{1}{2})(s_1{}^2 + s_2{}^2 + s_3{}^2)^{1/2}$ and being in the opposite states of polarization (χ,ψ) and $\left(-\chi, \psi + \frac{\pi}{2}\right)$, where χ and ψ are given by Eqs. (74).

Since we have used a fixed cartesian system of coordinates (x,y,z) to describe the Stokes parameters, the question of how these parameters change under a rotation of the axes naturally arises. To find the law of transformation, we need to consider only an elliptically polarized wave. This follows from the fact that a partially polarized wave always can be decomposed into two oppositely polarized independent waves. Let (s_0,s_1,s_2,s_3) denote the Stokes parameters of one of the elliptically polarized waves when referred to the original system, and let (s_0',s_1',s_2',s_3') denote these parameters when referred to the rotated system. The rotation consists of a clockwise twisting of the coordinates about the z axis and through an angle ϕ. By virtue of the fact that the wave is elliptically polarized, we can write the Stokes parameters (s_0,s_1,s_2,s_3) in terms of the ellipticity angle χ and orientation angle ψ, as follows:

$$s_0 \qquad s_0 \cos 2\chi \cos 2\psi \qquad s_0 \cos 2\chi \sin 2\psi \qquad s_0 \sin 2\chi$$

Obviously, when referred to the rotated system these parameters become

$$s_0 \qquad s_0 \cos 2\chi \cos 2(\psi - \phi) \qquad s_0 \cos 2\chi \sin 2(\psi - \phi) \qquad s_0 \sin 2\chi$$

Clearly then, the Stokes parameters referred to rotated coordinates are given by

$$\begin{aligned} s_0' &= s_0 \\ s_1' &= s_0 \cos 2\chi \cos 2(\psi - \phi) = s_1 \cos 2\phi + s_2 \sin 2\phi \\ s_2' &= s_0 \cos 2\chi \sin 2(\psi - \phi) = s_2 \cos 2\phi - s_1 \sin 2\phi \\ s_3' &= s_3 \end{aligned} \tag{75}$$

where (s_0,s_1,s_2,s_3) and (s_0',s_1',s_2',s_3') are the Stokes parameters in, respectively, the original and rotated coordinates. The parameters s_0 and s_3 are invariant under the rotation, i.e., the intensity and the ellipticity of the wave do not change when the axes are rotated. On the other hand, s_1 and s_2 do not remain the same and hence the orientation angle ψ changes when the axes are rotated.

5.4 Coherency Matrices

In the previous section it was demonstrated that the state of polarization of a narrowband polychromatic wave is specified completely by the four Stokes parameters s_0, s_1, s_2, s_3. In this section we shall show that the state of polarization can be specified alternatively by means of a 2×2 matrix whose elements characterize the state of coherency between the transverse components of the wave.

Let us again consider a plane TEM narrowband (quasi-monochromatic) polychromatic wave traveling in the z direction. In accord with Eqs. (42) the cartesian components of such a wave are

$$\begin{aligned} E_x &= \text{Re}\,\{a_x(t)e^{ikz}e^{-i\delta_x}e^{-i\omega t}\} \\ E_y &= \text{Re}\,\{a_y(t)e^{ikz}e^{-i\delta_y}e^{-i\omega t}\} \\ E_z &= 0 \end{aligned} \tag{76}$$

In terms of the complex vector **A**, whose components are given by

$$A_x = a_x(t)e^{ikz}e^{-i\delta_x} \qquad A_y = a_y(t)e^{ikz}e^{-i\delta_y} \qquad A_z = 0 \tag{77}$$

the electric field components (76) may be written in the form

$$E_x = \text{Re}\,\{A_x e^{-i\omega t}\} \qquad E_y = \text{Re}\,\{A_y e^{-i\omega t}\} \qquad E_z = 0 \tag{78}$$

which shows that **A** is the phasor of the electric vector of the wave. Unlike the phasor of a monochromatic wave, **A** is time-dependent.

The elements J_{pq} of the coherency matrix **J** are defined by

$$J_{pq} = \langle A_p A_q^* \rangle = \lim_{T\to\infty} \frac{1}{2T} \int_{-T}^{T} A_p A_q^*\, dt \qquad (p,q = x,y) \tag{79}$$

If A_p and A_q are physically independent then $\langle A_p A_q^* \rangle = 0$. It is obvious from definition (79) that

$$J_{xy} = J_{yx}^* \tag{80}$$

and hence the coherency matrix

$$\mathbf{J} = \begin{bmatrix} J_{xx} & J_{xy} \\ J_{yx} & J_{yy} \end{bmatrix} = \begin{bmatrix} \langle A_x A_x^* \rangle & \langle A_x A_y^* \rangle \\ \langle A_y A_x^* \rangle & \langle A_y A_y^* \rangle \end{bmatrix} \tag{81}$$

is hermitian.

To find the connection between the Stokes parameters and the coherency matrix, we note that when expressions (77) are substituted into definition (79) we get

$$\begin{aligned} J_{xx} &= \langle a_x^2(t) \rangle \qquad J_{yy} = \langle a_y^2(t) \rangle \\ J_{xy} &= \langle a_x(t)a_y(t)e^{i\delta(t)} \rangle = \langle a_x(t)a_y(t)\cos\delta(t) \rangle + i\langle a_x(t)a_y(t)\sin\delta(t) \rangle \\ J_{yx} &= \langle a_x(t)a_y e^{-i\delta(t)} \rangle = \langle a_x(t)a_y(t)\cos\delta(t) \rangle - i\langle a_x(t)a_y(t)\sin\delta(t) \rangle \end{aligned} \tag{82}$$

where $\delta(t) \equiv \delta_y(t) - \delta_x(t)$. Comparing expressions (82) with expressions (43), we find that the Stokes parameters are related to the elements of the coherency matrix as follows:

$$\begin{aligned} s_0 &= J_{xx} + J_{yy} & s_1 &= J_{xx} - J_{yy} \\ s_2 &= J_{xy} + J_{yx} & s_3 &= i(J_{yx} - J_{xy}) \\ J_{xx} &= \tfrac{1}{2}(s_0 + s_1) & J_{yy} &= \tfrac{1}{2}(s_0 - s_1) \\ J_{xy} &= \tfrac{1}{2}(s_2 + is_3) & J_{yx} &= \tfrac{1}{2}(s_2 - is_3) \end{aligned} \tag{83}$$

These relations show that the Stokes parameters and the elements of the coherency matrix are linearly related and that a specification of the wave in terms of the latter is in all respects equivalent to its specification in terms of the former.[1]

Since the additivity theorem applies to the Stokes parameters, it must, in view of the linear relations (83), also apply to the coherency matrix, in the sense that if $\mathbf{J}^{(1)}, \mathbf{J}^{(2)}, \ldots, \mathbf{J}^{(N)}$ are the coherency matrices of N independent waves traveling in the same direction, then the coherency matrix $\mathbf{J}$ of the resulting wave is the sum of the coherency matrices of the independent waves, viz.,

$$\mathbf{J} = \sum_{n=1}^{N} \mathbf{J}^{(n)} \tag{84}$$

To show this, we let $A_x^{(n)}$, $A_y^{(n)}$ be the cartesian components of the phasor of the nth independent wave. Then by superposition the cartesian components of the phasor of the resulting wave are

$$A_x = \sum_{n=1}^{N} A_x^{(n)} \qquad A_y = \sum_{n=1}^{N} A_y^{(n)} \tag{85}$$

The elements of the coherency matrix of the resulting wave are

$$\begin{aligned} J_{pq} = \langle A_p A_q^* \rangle &= \sum_{n=1}^{N} \sum_{m=1}^{N} \langle A_p^{(n)} A_q^{(m)*} \rangle \\ &= \sum_{n=1}^{N} \langle A_p^{(n)} A_q^{(n)*} \rangle + \sum_{n \neq m} \langle A_p^{(n)} A_q^{(m)*} \rangle \end{aligned} \tag{86}$$

[1] Compare with E. Wolf, Coherence Properties of Partially Polarized Electromagnetic Radiation, *Nuovo Cimento,* **13:** 1165 (1959).

Each term of the last summation is zero since $A_p^{(n)}$ and $A_q^{(m)}$ for $n \neq m$ are independent. Hence we have

$$J_{pq} = \sum_{n=1}^{N} J_{pq}^{(n)} \tag{87}$$

where $J_{pq}^{(n)}$ denotes the elements of the coherency matrix of the nth independent wave. Thus the additivity theorem (84) is verified.

From Schwarz's inequality, which is expressed by

$$\int A_p A_p^* \, dt \int A_q A_q^* \, dt \geq \int A_p^* A_q \, dt \int A_p A_q^* \, dt$$

and from definition (79) it follows that

$$J_{xx} J_{yy} \geq J_{xy}^* J_{xy} \tag{88}$$

or, because of Eq. (80), that

$$J_{xx} J_{yy} - J_{yx} J_{xy} \geq 0 \tag{89}$$

The equality sign in these expressions obtains only when A_p/A_q is constant, which in turn means that the determinant of the coherency matrix vanishes only if the wave is elliptically polarized. If the determinant does not vanish, then the wave is partially polarized. That is,

$$\begin{aligned} \det \mathbf{J} &= 0 \qquad \text{for elliptic polarization} \\ \det \mathbf{J} &> 0 \qquad \text{for partial polarization} \end{aligned} \tag{90}$$

We know from our study of the Stokes parameters that for an unpolarized wave $s_0 \neq 0$ and $s_1 = s_2 = s_3 = 0$. Casting this into the language of the coherency matrix, we see from Eqs. (83) that $J_{xx} = J_{yy} = (\frac{1}{2})s_0$. Thus we find that the coherency matrix of an unpolarized wave has the form

$$\mathbf{J} = \frac{s_0}{2} \begin{bmatrix} 1 & 0 \\ 0 & 1 \end{bmatrix} \tag{91}$$

Moreover, from expressions (46) and Eqs. (83) we see that the coherency matrix of an elliptically polarized wave has the form

$$\mathbf{J} = \frac{s_0}{2}\begin{bmatrix} (1 + \cos 2\chi \cos 2\psi) & (\cos 2\chi \sin 2\psi + i \sin 2\chi) \\ (\cos 2\chi \sin 2\psi - i \sin 2\chi) & (1 - \cos 2\chi \cos 2\psi) \end{bmatrix} \tag{92}$$

where ψ is the orientation angle of the polarization ellipse and χ is its ellipticity angle. To see what this matrix looks like for certain simple states of polarization, we recall that for linear polarization $\chi = 0$, for right-handed circular polarization $\chi = -\pi/4$, and for left-handed circular polarization $\chi = \pi/4$. Hence from expression (92) we find that

$$\mathbf{J} = \frac{s_0}{2}\begin{bmatrix} 1 + \cos 2\psi & \sin 2\psi \\ \sin 2\psi & 1 - \cos 2\psi \end{bmatrix} \tag{93}$$

is the coherency matrix of a linearly polarized wave making an angle ψ with the x axis;

$$\mathbf{J} = \frac{s_0}{2}\begin{bmatrix} 1 & -i \\ i & 1 \end{bmatrix} \tag{94}$$

is the coherency matrix for right-circular polarization; and

$$\mathbf{J} = \frac{s_0}{2}\begin{bmatrix} 1 & i \\ -i & 1 \end{bmatrix} \tag{95}$$

is the coherency matrix for left-circular polarization.

It follows from relations (83) that the coherency matrix can be expanded in terms of the Stokes parameters and certain elementary matrices which in wave mechanics are called the Pauli spin matrices.[1] That is,

$$\mathbf{J} = \frac{1}{2}\sum_{p=0}^{3} s_p \boldsymbol{\sigma}_p \tag{96}$$

where s_p $(p = 0, 1, 2, 3)$ are the Stokes parameters, $\boldsymbol{\sigma}_0$ is the unit

[1] See, for example, U. Fano, A Stokes-parameter Technique for the Treatment of Polarization in Quantum Mechanics, *Phys. Rev.*, **93:** 121 (1954).

matrix

$$\boldsymbol{\sigma}_0 = \begin{bmatrix} 1 & 0 \\ 0 & 1 \end{bmatrix} \tag{97}$$

and $\boldsymbol{\sigma}_1$, $\boldsymbol{\sigma}_2$, $\boldsymbol{\sigma}_3$ are the Pauli spin matrices

$$\boldsymbol{\sigma}_1 = \begin{bmatrix} 1 & 0 \\ 0 & -1 \end{bmatrix} \qquad \boldsymbol{\sigma}_2 = \begin{bmatrix} 0 & 1 \\ 1 & 0 \end{bmatrix} \qquad \boldsymbol{\sigma}_3 = \begin{bmatrix} 0 & i \\ -i & 0 \end{bmatrix} \tag{98}$$

From Eq. (91) we see that $\boldsymbol{\sigma}_0$ represents an unpolarized wave. Furthermore, using the decompositions

$$\begin{aligned} \boldsymbol{\sigma}_1 &= \begin{bmatrix} 1 & 0 \\ 0 & -1 \end{bmatrix} = \begin{bmatrix} 1 & 0 \\ 0 & 0 \end{bmatrix} - \begin{bmatrix} 0 & 0 \\ 0 & 1 \end{bmatrix} \\ \boldsymbol{\sigma}_2 &= \begin{bmatrix} 0 & 1 \\ 1 & 0 \end{bmatrix} = \tfrac{1}{2}\begin{bmatrix} 1 & 1 \\ 1 & 1 \end{bmatrix} - \tfrac{1}{2}\begin{bmatrix} 1 & -1 \\ -1 & 1 \end{bmatrix} \\ \boldsymbol{\sigma}_3 &= \begin{bmatrix} 0 & i \\ -i & 0 \end{bmatrix} = \tfrac{1}{2}\begin{bmatrix} 1 & i \\ -i & 1 \end{bmatrix} - \tfrac{1}{2}\begin{bmatrix} 1 & -i \\ i & 1 \end{bmatrix} \end{aligned} \tag{99}$$

and recalling the states of polarization that the matrices (93), (94), and (95) express, we see that $\boldsymbol{\sigma}_1$ characterizes the excess of a linearly polarized wave making an angle $\psi = 0$ over a linearly polarized wave making an angle $\psi = \pi/2$; $\boldsymbol{\sigma}_2$ the excess of a linearly polarized wave making an angle $\psi = \pi/4$ over a linearly polarized wave making an angle $\psi = 3\pi/4$; and $\boldsymbol{\sigma}_3$ the excess of a wave polarized circularly to the left over one polarized circularly to the right.

If we decompose the wave into an unpolarized part and an elliptically polarized part, then the ratio of the intensity of the polarized part to the intensity of the original wave is the degree of polarization m of the wave. The quantity

$$\operatorname{Tr} \mathbf{J} \equiv J_{xx} + J_{yy} \tag{100}$$

is the trace (or spur) of the matrix and represents the intensity of the original wave. The degree of polarization is given by the expression

$$m = \sqrt{1 - 4 \det \mathbf{J}/(\operatorname{Tr} \mathbf{J})^2} \tag{101}$$

which can be derived from Eqs. (62) and (83). Since this expression involves only the rotational invariants det **J** and Tr **J**, the degree of polarization does not change with a rotation of the coordinate axis. From Eqs. (63) and (83) it follows that the orientation of the polarization ellipse of the polarized part of the wave is given by

$$\tan 2\psi = \frac{J_{xy} + J_{yx}}{J_{xx} - J_{yy}} \tag{102}$$

and from Eqs. (64) and (83) that its ellipticity is given by

$$\sin 2\chi = i\,\frac{J_{yx} - J_{xy}}{\sqrt{(\mathrm{Tr}\,\mathbf{J})^2 - 4\det\mathbf{J}}} \tag{103}$$

Under rotation χ does not change because the denominator of Eq. (103) is a rotational invariant, as is the numerator $i(J_{yx} - J_{xy})$. However, ψ does change under rotation, as might have been expected. Thus we see that m and χ are independent of the choice of orientation of the coordinate axes, while ψ is not.

The quantities Tr **J** and det **J** do not change when the coherency matrix is transposed; on the other hand, the quantity $J_{yx} - J_{xy}$ simply changes in sign. Therefore, from Eq. (103) we see that χ simply changes in sign when the coherency matrix is transposed. Since the sign of χ determines the sense of polarization, this means that if a coherency matrix describes a wave with a certain sense of polarization, then the transpose of the matrix describes a wave traveling in the same direction but with the opposite sense of polarization; or if a coherency matrix describes a wave traveling in a certain direction, the same matrix also describes a wave traveling in the opposite direction with opposite polarization.

5.5 Reception of Partially Polarized Waves

In this section we shall calculate how much power a given antenna can extract from an incident polychromatic wave. We shall carry out the calculation by recalling the results of the conventional case, where the incoming wave is monochromatic, and then generalizing these results to the case where the incoming wave is polychromatic. This method of analyzing the problem, which uses the monochromatic theory of antennas as a point of departure, appears to be the most tractable, because it takes advantage of the fact that the receiving properties of an antenna are most conveniently expressed in terms of its monochromatic behavior as a transmitter.

Hence, for the present we confine our attention to the conventional monochromatic theory of receiving antennas. According to this theory an antenna, actually or effectively, has two circuit terminals and with respect to these terminals its behavior is as follows: When the antenna is driven by a monochromatic voltage source applied to its terminals and no radiation is incident, the source "sees" an impedance, namely, the input impedance Z_i of the antenna; on the other hand, when a monochromatic wave of the same frequency is incident on the antenna and the terminals are open-circuited, a voltage appears across the terminals, namely, the open-circuit voltage V_0. Then, in accord with Thévenin's theorem of circuit theory, when the antenna operates as a receiving antenna having a load impedance Z_l connected to its terminals, the equivalent circuit of the antenna consists of the voltage V_0 in series with Z_i and Z_l. From this equivalent circuit it is clear that the power absorbed by the load is a maximum when Z_i and Z_l are conjugate-matched, i.e., $Z_i = Z_l^*$. Under this condition of optimum power transfer, the power generated by V_0 is divided equally between the power absorbed by Z_i and the power absorbed by Z_l. Physically, the power absorbed by Z_i consists of the (reversible) power that is carried away from the antenna by the scattered, or reradiated, portion of the incident power and the (irreversible) power that goes into ohmic losses, i.e., into the heating of the antenna structure. In the hypothetical case where the conjugate-matched antenna is free of ohmic losses, one-half of the applied incident power is scattered into space and the other half is absorbed by the load.

The power that an incident monochromatic wave delivers to the conjugate-matched load of a receiving antenna is related to the behavior of the antenna as a transmitter. To present this relation, let us suppose that the antenna in question is driven as a transmitter by a monochromatic voltage applied to its terminals. Let us also suppose that the antenna is located at the origin of a spherical coordinate system

(r,θ,ϕ). Then if the electric vector (actually phasor) of the far-zone field radiated by the antenna is $\mathbf{E}^{rad}$, the radial component of the Poynting vector of this field is

$$S^{rad}(r,\theta,\phi) = \frac{1}{2}\sqrt{\frac{\epsilon_0}{\mu_0}}\,\mathbf{E}^{rad}\cdot\mathbf{E}^{rad*} \tag{104}$$

the field polarization vector is

$$\mathbf{p}^{rad}(\theta,\phi) = \frac{\mathbf{E}^{rad}}{\sqrt{\mathbf{E}^{rad}\cdot\mathbf{E}^{rad*}}} \tag{105}$$

and the gain function is

$$g(\theta,\phi) = \frac{4\pi r^2 S^{rad}(r,\theta,\phi)}{\int_0^{4\pi} S^{rad}(r,\theta,\phi) r^2\, d\Omega} \tag{106}$$

where $d\Omega(= \sin\theta\, d\theta\, d\phi)$ is an element of solid angle. Alternatively, let us now suppose that the antenna is operated as a receiving antenna with a conjugate-matched load attached to its terminals, and that a plane monochromatic wave is incident on it from a direction $\theta = \theta_0$, $\phi = \phi_0$. If the electric vector of the incident wave is $\mathbf{E}^{inc}$, the radial component of the Poynting vector of the incident wave is

$$S^{inc}(\theta_0,\phi_0) = \frac{1}{2}\sqrt{\frac{\epsilon_0}{\mu_0}}\,\mathbf{E}^{inc}\cdot\mathbf{E}^{inc*} \tag{107}$$

and the field polarization vector is

$$\mathbf{p}^{inc}(\theta_0,\phi_0) = \frac{\mathbf{E}^{inc}}{\sqrt{\mathbf{E}^{inc}\cdot\mathbf{E}^{inc*}}} \tag{108}$$

Then, in compliance with the reciprocity theorem,[1] the power absorbed by the load is given by the relation

$$P_{abs} = \frac{\lambda^2}{4\pi}\, g(\theta_0,\phi_0) S^{inc}(\theta_0,\phi_0)\,|\mathbf{p}^{rad}(\theta_0,\phi_0)\cdot\mathbf{p}^{inc}(\theta_0,\phi_0)|^2 \tag{109}$$

[1] S. A. Schelkunoff and H. T. Friis, "Antennas: Theory and Practice," pp. 390–394, John Wiley & Sons, Inc., New York, 1952.

where the quantities $g(\theta,\phi)$ and $\mathbf{p}^{rad}(\theta,\phi)$ describe the behavior of the antenna in transmission and the quantities $S^{inc}(\theta,\phi)$ and $\mathbf{p}^{inc}(\theta,\phi)$ describe the incident wave in reception.

The polarization loss factor

$$K(\theta,\phi) = |\mathbf{p}^{rad}(\theta,\phi) \cdot \mathbf{p}^{inc}(\theta,\phi)|^2 \tag{110}$$

which appears in relation (109) can take on any value in the range $0 \leq K \leq 1$, depending on how closely the polarization of the wave radiated in a direction (θ,ϕ) is matched to the polarization of the incident wave falling on the antenna from the same direction. When

$$\mathbf{p}^{rad} = \mathbf{p}^{inc*} \tag{111}$$

the radiated wave and the incident wave are matched completely and $K = 1$. If the field polarization vector of the incident wave is conjugate-matched in this sense to the field polarization vector of the radiated wave, the power absorbed by the conjugate-matched load is a maximum and, according to Eq. (109), has the value[1]

$$(P_{abs})_{\max} = \frac{\lambda^2}{4\pi} g(\theta,\phi) S^{inc}(\theta,\phi) \tag{112}$$

By definition the ratio $(P_{abs})_{\max}/S^{inc}(\theta,\phi)$ is the effective area $A(\theta,\phi)$ of the receiving antenna,[2] and consequently the effective area of the antenna in reception is proportional to the gain function of the antenna in transmission, i.e.,

$$A(\theta,\phi) = \frac{\lambda^2}{4\pi} g(\theta,\phi) \tag{113}$$

With the aid of this result and definition (110) we can write Eq. (109) in the alternative form

$$P_{abs} = A(\theta,\phi) S^{inc}(\theta,\phi) K(\theta,\phi) \tag{114}$$

[1] Y.-C. Yeh, The Received Power of a Receiving Antenna and the Criteria for Its Design, *Proc. IRE*, **37:** 155 (1949).

[2] Compare C. T. Tai, On the Definition of the Effective Aperture of Antennas, *IRE Trans. Antennas Propagation*, **AP-9:** 224–225 (March, 1961).

which explicitly displays the dependence of the absorbed power on the effective area of the antenna and on the polarization loss factor.

To generalize the above discussion to the case where the incident wave is partially polarized and polychromatic, we write Eq. (114) in the equivalent form

$$P_{abs} = A(\theta,\phi)S^{inc}\ \mathrm{Tr}\ (\mathbf{p}^{rad}\mathbf{p}^{rad*}) \cdot (\widetilde{\mathbf{p}^{inc}\mathbf{p}^{inc*}}) \tag{115}$$

where $\mathbf{p}^{rad}\mathbf{p}^{rad*}$ is the dyadic associated with the wave radiated in a direction (θ,ϕ), and $\widetilde{\mathbf{p}^{inc}\mathbf{p}^{inc*}}$ denotes the transpose of the dyadic $\mathbf{p}^{inc}\mathbf{p}^{inc*}$ associated with the wave incident from the same direction (θ,ϕ). Moreover, we can in turn write Eq. (115) as

$$P_{abs} = \frac{1}{2}\sqrt{\frac{\epsilon_0}{\mu_0}}\,A(\theta,\phi)\ \mathrm{Tr}\ (\mathbf{p}^{rad}\mathbf{p}^{rad*}) \cdot (\widetilde{\mathbf{E}^{inc}\mathbf{E}^{inc*}}) \tag{116}$$

Now if the incident wave happens to be a polychromatic wave and if over the entire spectrum of the wave the antenna is conjugate-matched to the load, then Eq. (116) remains valid for each frequency of the spectrum. Assuming that the antenna and load are so matched, we get the total absorbed power by integrating Eq. (116) over all frequencies or, as mentioned in Sec. 5.1, by averaging with respect to time. Such an integration would require a knowledge of the frequency dependence of $\mathbf{p}^{rad}$ and A. However, we shall assume that $\mathbf{p}^{rad}$, and hence A, is independent of frequency over the spectrum of the polychromatic wave and shall thus obtain the following expression for the total absorbed power

$$P_{abs} = \frac{1}{2}\sqrt{\frac{\epsilon_0}{\mu_0}}\,A(\theta,\phi)\ \mathrm{Tr}\ (\mathbf{p}^{rad}\mathbf{p}^{rad*}) \cdot (\widetilde{\langle\mathbf{E}^{inc}\mathbf{E}^{inc*}\rangle}) \tag{117}$$

Let us now consider the case where the incident polychromatic wave is narrowband and has the form

$$\mathbf{E}^{inc}(\mathbf{r},t) = [\mathbf{e}_\theta E_\theta(t) + \mathbf{e}_\phi E_\phi(t)]e^{-ikr}e^{-i\omega t} \tag{118}$$

Here the complex components $E_\theta(t)$ and $E_\phi(t)$ are slowly varying functions of time, ω is a mean frequency, and $k = \omega/c$. For such an incident wave the matrix of the components of the time-average value of the dyadic $\mathbf{E}^{inc}\mathbf{E}^{inc*}$ is

$$[I_{ij}] = \frac{1}{2}\sqrt{\frac{\epsilon_0}{\mu_0}}\begin{bmatrix}\langle E_\theta E_\theta^*\rangle & \langle E_\theta E_\phi^*\rangle \\ \langle E_\phi E_\theta^*\rangle & \langle E_\phi E_\phi^*\rangle\end{bmatrix} \qquad (i = 1, 2; j = 1, 2) \tag{119}$$

The coherency matrix of the incident wave in Eq. (117) is the transpose $[\widetilde{I_{ij}}]$ of $[I_{ij}]$. The matrix of the components of the dyadic $A(\mathbf{p}^{rad}\mathbf{p}^{rad*})$, that is,

$$[A_{ij}] = A(\theta,\phi)\begin{bmatrix}p_\theta{}^{rad}p_\theta{}^{rad*} & p_\theta{}^{rad}p_\phi{}^{rad*} \\ p_\phi{}^{rad}p_\theta{}^{rad*} & p_\phi{}^{rad}p_\phi{}^{rad*}\end{bmatrix} \qquad (i = 1, 2; j = 1, 2) \tag{120}$$

is the effective-area matrix of the antenna. In terms of the effective-area matrix $[A_{ij}]$ of the receiving antenna and the coherency matrix $[\widetilde{I_{ij}}]$ of the incident wave, the power absorbed in the conjugate-matched load of the receiving antenna is given by the compact relation

$$P_{abs} = \mathrm{Tr}\,[A_{ij}][\widetilde{I_{ij}}] \qquad (i = 1, 2; j = 1, 2) \tag{121}$$

which follows directly from Eq. (116) and definitions (119), (120).

We can divide the incident wave into two mutually independent parts, viz., an unpolarized part and a polarized part. We do this by splitting $[I_{ij}]$ into

$$\begin{bmatrix}I_{11} & I_{12} \\ I_{21} & I_{22}\end{bmatrix} = \alpha\begin{bmatrix}1 & 0 \\ 0 & 1\end{bmatrix} + \beta\begin{bmatrix}q_{11} & q_{12} \\ q_{21} & q_{22}\end{bmatrix} \tag{122}$$

and noting that the first matrix on the right represents the unpolarized part and the second matrix on the right represents the polarized part. Taking the trace of this matrix equation, we obtain the expression

$$I_{11} + I_{22} = 2\alpha + \beta(q_{11} + q_{22}) \tag{123}$$

whose left side represents the average value $\langle S^{inc}\rangle$ of the incident power density, and whose right side represents the power density 2α of its unpolarized part plus the power density $\beta(q_{11} + q_{22})$ of its polarized part. By definition the degree of polarization m is the ratio of the power density of the polarized part to the total power density; hence in terms of m Eq. (123) can be written as

$$I_{11} + I_{22} = 2\alpha + m(I_{11} + I_{22}) \tag{124}$$

From this it follows that

$$\alpha = \tfrac{1}{2}(1 - m)(I_{11} + I_{22}) = \tfrac{1}{2}(1 - m)\langle S^{inc}\rangle \tag{125}$$

Since we are free to choose β, we make the choice

$$\beta = m\langle S^{inc}\rangle \tag{126}$$

on the grounds of convenience. In view of expressions (125) and (126) we see that Eq. (122) can be written in terms of the time-average power density $\langle S^{inc}\rangle$ of the incident wave and its degree m of polarization:

$$\begin{bmatrix} I_{11} & I_{12} \\ I_{21} & I_{22} \end{bmatrix} = \tfrac{1}{2}(1 - m)\langle S^{inc}\rangle \begin{bmatrix} 1 & 0 \\ 0 & 1 \end{bmatrix} + m\langle S^{inc}\rangle \begin{bmatrix} q_{11} & q_{12} \\ q_{21} & q_{22} \end{bmatrix} \tag{127}$$

As a consequence of choice (126) we have

$$q_{11} + q_{22} = 1 \tag{128}$$

and because $[q_{ij}]$ represents a completely polarized wave, we have

$$q_{11}q_{22} - q_{12}q_{21} = 0 \tag{129}$$

Moreover, by virtue of the fact that $[I_{ij}]$ is hermitian, we also have

$$q_{12} = q_{21}^* \tag{130}$$

From conditions (128), (129), and (130) we see that the components of q_{ij} may be written in the following way in terms of the orientation angle ψ and the ellipticity χ of the polarization ellipse of the polarized part of the incident wave falling on the antenna from a direction (θ,ϕ):

$$\begin{bmatrix} q_{11} & q_{12} \\ q_{21} & q_{22} \end{bmatrix} = \tfrac{1}{2} \begin{bmatrix} 1 + \cos 2\chi \cos 2\psi & \cos 2\chi \sin 2\psi + i \sin 2\chi \\ \cos 2\chi \sin 2\psi - i \sin 2\chi & 1 - \cos 2\chi \cos 2\psi \end{bmatrix} \tag{131}$$

Similarly, since $[A_{ij}]$ represents a completely polarized wave, viz., the wave the antenna would radiate if it were used as a transmitter, we can write it in terms of the orientation angle ψ' and ellipticity angle χ' of the wave radiated in a direction (θ,ϕ):

$$[A_{ij}] = \tfrac{1}{2}A(\theta,\phi)\begin{bmatrix} 1 + \cos 2\chi' \cos 2\psi' & \cos 2\chi' \sin 2\psi' + i \sin 2\chi' \\ \cos 2\chi' \sin 2\psi' - i \sin 2\chi' & 1 - \cos 2\chi' \cos 2\psi' \end{bmatrix} \tag{132}$$

Substituting Eq. (127) into Eq. (121), we get

$$P_{abs} = \operatorname{Tr}\,[A_{ij}][\widetilde{I_{ij}}] = \tfrac{1}{2}(1 - m)(A_{11} + A_{22})\langle S^{inc}\rangle + m(A_{11}q_{11} + A_{12}q_{12} + A_{21}q_{21} + A_{22}q_{22})\langle S^{inc}\rangle \tag{133}$$

and then, using Eqs. (131) and (132), we find that the time-average power absorbed by the conjugate-matched load is given by[1]

$$P_{abs} = \tfrac{1}{2}(1 - m)A(\theta,\phi)\langle S^{inc}(\theta,\phi)\rangle + mA(\theta,\phi)\langle S^{inc}(\theta,\phi)\rangle \cos^2 \frac{\gamma}{2} \tag{134}$$

where

$$\cos \gamma \equiv \cos 2\chi' \cos 2\chi \cos (2\psi' - 2\psi) - \sin 2\chi' \sin 2\chi \tag{135}$$

On the Poincaré sphere, γ is the angle between the point $(2\psi,-2\chi)$ describing the polarization ellipse of the incident wave and the point $(2\psi',2\chi')$ describing the polarization ellipse of the radiated wave. When $\psi' = \psi$ and $\chi' = -\chi$, that is, the two points coincide and $\gamma = 0$, the polarizations of the radiated and incident waves are conjugate-matched and there is no polarization loss. This, of course, means that the two polarization ellipses have the same orientation in space and the same axial ratio. It also means that the sense of rotation of the incident wave is the same as the sense of rotation of the radiated wave if the former is viewed from infinity and the latter from the antenna. If viewed from some fixed position, the senses of rotation would appear to be opposite.

[1] H. C. Ko, Theoretical Techniques for Handling Partially Polarized Radio Waves with Special Reference to Antennas, *Proc. IRE*, **49:** 1446 (1961).

The first term on the right of Eq. (134) represents the contribution to P_{abs} of the unpolarized part of the incident wave, whereas the second represents the contribution of the polarized part. If the polarization of the antenna in a direction (θ,ϕ) is conjugate-matched to the incident wave coming from the same direction, then $\gamma = 0$ and the power absorbed in the conjugate-matched load resistance is a maximum, i.e.,

$$(P_{abs})_{\max} = \tfrac{1}{2}(1 + m)A(\theta,\phi)\langle S^{inc}(\theta,\phi)\rangle \tag{136}$$

Moreover, if the incident wave is completely polarized, we have $m = 1$ and hence

$$(P_{abs})_{\max} = A(\theta,\phi)\langle S^{inc}(\theta,\phi)\rangle \tag{137}$$

On the other hand, if the incident wave is completely unpolarized, we have $m = 0$ and hence

$$P_{abs} = \tfrac{1}{2}A(\theta,\phi)\langle S^{inc}(\theta,\phi)\rangle \tag{138}$$

In this case there is no question of matching.

5.6 Antenna Temperature and Integral Equation for Brightness Temperature

From the discussion in the previous section we know that if a plane unpolarized polychromatic wave is incident from a direction θ, ϕ on a lossless receiving antenna located at the origin of a spherical coordinate system (r,θ,ϕ), the power absorbed by the matched load (the receiver) is given by

$$P_{abs} = \tfrac{1}{2}A(\theta,\phi)S^{inc}(\theta,\phi)\Delta\omega \tag{139}$$

where S^{inc} is the spectral flux density of the incident wave, $\Delta\omega$ the bandwidth of the receiver, and A the effective area of the antenna. The validity of this expression rests on the assumption that A and S^{inc} are independent of frequency within the relatively narrow bandwidth

$\Delta\omega$. To find the absorbed power for the case where the source is distributed over the sky, we note that the elemental contribution to the absorbed power of the radiation falling within a cone of solid angle $d\Omega$ and within a bandwidth $\Delta\omega$ can be expressed as

$$dP_{abs} = \tfrac{1}{2}A(\theta,\phi)dS^{inc}(\theta,\phi)\Delta\omega \tag{140}$$

where, in view of Eq. (23), dS^{inc} is related to the brightness temperature T_b of the sky by

$$dS^{inc}(\theta,\phi) = \frac{2k}{\lambda^2} T_b(\theta,\phi)d\Omega \tag{141}$$

The subscript ω has been dropped from T_b for simplicity. Then we assume that the radiation falling on the antenna from any direction is incoherent with respect to the radiation from the other directions. By virtue of this assumption, the total absorbed power is the sum of the elemental powers delivered to the antenna by various incident rays. In other words, if the incident rays are physically independent, the total absorbed noiselike power may be calculated by integrating expression (140) over the solid angle subtended by the distributed source:

$$P_{abs} = k\,\Delta\omega\left[\frac{1}{\lambda^2}\int A(\theta,\phi)T_b(\theta,\phi)d\Omega\right] \tag{142}$$

The quantity in the brackets has the dimension of temperature and is known as the antenna temperature. It provides a convenient measure for the noiselike power picked up by the antenna in a bandwidth $\Delta\omega$. Thus antenna temperature T_a is defined by

$$T_a = \frac{1}{\lambda^2}\int A(\theta,\phi)T_b(\theta,\phi)d\Omega \tag{143}$$

or, in terms of the gain function $g(\theta,\phi)$, by

$$T_a = \frac{1}{4\pi}\int g(\theta,\phi)T_b(\theta,\phi)d\Omega \tag{144}$$

According to this definition, one possible physical interpretation of T_a is as follows: If the antenna is completely enclosed by a surface which radiates as a blackbody at temperature T_a, then the antenna will absorb in its load resistance the power $kT_a\,\Delta\omega$. Alternatively, T_a may be regarded as the temperature to which the effective input resistance of the receiver (which, if matched, equals the radiation resistance of the antenna) must be raised so that the noise power, produced by the thermal motion of the electrons and delivered to the receiver through a lossless line, would equal P_{abs} in accordance with the relation $P_{abs} = kT_a\,\Delta\omega$. Antenna temperature as defined by Eq. (144) is a measure of the incident radiation only; it is not a measure of the temperature of the material in the antenna structure.

So far we have tacitly assumed that the direction of the main lobe of the receiving antenna is fixed and lies along the axis $\theta = 0$ of the spherical coordinate system. However, this is an unnecessary restriction and can be removed easily. For example, if we let $\mathbf{n}$ be a unit vector pointing in the direction of the main lobe and $\mathbf{n}'$ be a unit vector in the direction of the solid angle $d\Omega(\mathbf{n}')$, then Eq. (144) can be formally written in the following more general form:

$$T_a(\mathbf{n}) = \frac{1}{4\pi}\int g(\mathbf{n},\mathbf{n}')T_b(\mathbf{n}')d\Omega(\mathbf{n}') \tag{145}$$

This is the integral equation for the brightness temperature $T_b(\mathbf{n}')$. By changing the orientation $\mathbf{n}$ of the antenna so that its radiation pattern effectively scans the sky, we can measure T_a as a function of $\mathbf{n}$. Moreover, by measuring the radiation pattern or by predicting it theoretically, we can deduce the gain function. Accordingly, we regard $T_a(\mathbf{n})$ and $g(\mathbf{n},\mathbf{n}')$ as known quantities, and find the brightness temperature $T_b(\mathbf{n}')$ of the sky in terms of $T_a(\mathbf{n})$ and $g(\mathbf{n},\mathbf{n}')$ by solving the integral equation. A practical way of solving the integral equation is by successive approximations.[1] To show what the scheme of the method is, let us write Eq. (145) in operator form

$$T_a(\mathbf{n}) = K(\mathbf{n},\mathbf{n}')T_b(\mathbf{n}') \tag{146}$$

[1] See J. G. Bolton and K. C. Westfold, Galactic Radiation at Radio Frequencies, *Australian J. Sci. Res.*, **3:** 19 (1950).

where $K(\mathbf{n},\mathbf{n}')$ is the integral operator defined by

$$K(\mathbf{n},\mathbf{n}')f(\mathbf{n}') \equiv \frac{1}{4\pi}\int g(\mathbf{n},\mathbf{n}')f(\mathbf{n}')d\Omega(\mathbf{n}') \tag{147}$$

the quantity $f(\mathbf{n}')$ being a typical function of $\mathbf{n}'$. Also, for simplicity, we do not bother to write explicitly the arguments $\mathbf{n}$, $\mathbf{n}'$. Thus in operator form Eq. (145) becomes

$$T_a = KT_b$$

or equivalently

$$T_b = T_a + (1 - K)T_b \tag{148}$$

Suppose as a zero-order approximation to T_b we choose the known function T_a and then take

$$T_1 = T_a + (1 - K)T_a \tag{149}$$

as the first-order approximation to T_b. By applying the same procedure to T_1, we obtain the second-order approximation to T_b:

$$T_2 = T_a + (1 - K)T_1 \tag{150}$$

Clearly for the nth approximation to T_b we have

$$T_n = T_a + (1 - K)T_{n-1} \tag{151}$$

or in terms of T_a

$$T_n = T_a + (1 - K)T_a + (1 - K)^2T_a + \cdots (1 - K)^nT_a \tag{152}$$

5.7 Elementary Theory of the Two-element Radio Interferometer

To attain high resolving power, antenna arrays having multilobe receiving patterns are used. The high resolving power of such arrays stems

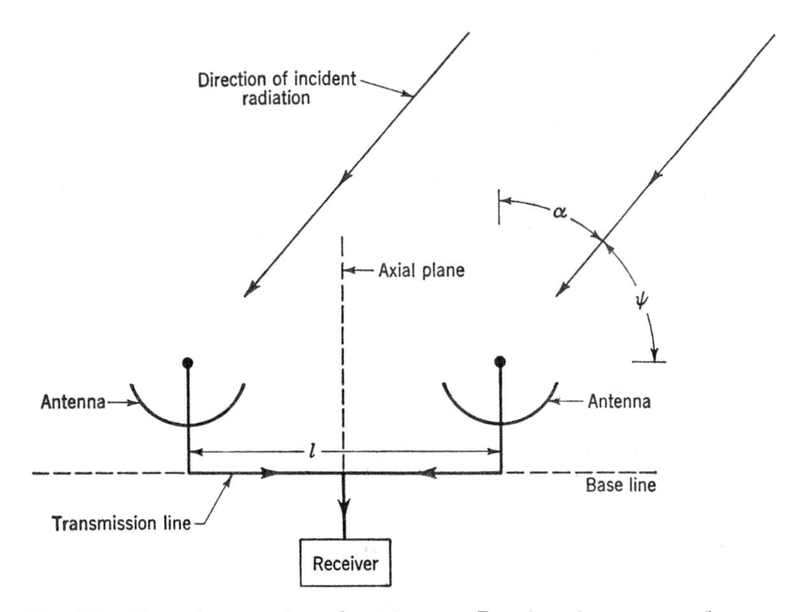

Fig. 5.7 ***Two-element interferometer. Receiver is connected to two identical and similarly oriented antennas. Direction of incident radiation makes angle ψ with base line and angle α with axial plane. Separation of antennas is l. Receiver is at electrical center of transmission line.***

from the fact that each lobe of the multilobe pattern becomes narrower and hence more resolvent as the spacing between adjacent antennas is increased.

The simplest array that exhibits a multilobe receiving pattern is the two-element radio interferometer,[1] consisting of two identical and similarly oriented receiving antennas separated by a distance l and connected to a single tuned[2] receiver by a transmission line (Fig. 5.7). To find the receiving pattern of such an interferometer, we note that by

[1] One of the first applications of the two-element radio interferometer, which we recognize as the radio analog of Michelson's optical interferometer, was made by L. L. McReady, J. L. Pawsey, and R. Payne-Scott, Solar Radiation at Radio Frequencies and Its Relation to Sunspots, *Proc. Roy. Soc.*, (A) **190:** 357 (1947).

[2] Because the receiver is sharply tuned we can use a monochromatic theory in most of the analysis.

the reciprocity theorem its receiving and radiation patterns must be the same and we recall from Sec. 3.5 that its radiation pattern must be the product of the radiation pattern F of one antenna and the array factor A of the two antennas. Thus it follows from the reciprocity and multiplication theorems that the receiving pattern of the interferometer is $|FA|$ and that the power fed to the receiver is proportional to $|FA|^2$. In a typical two-element interferometer F has one main lobe and the factor A has numerous lobes; these are called "grating" lobes. Consequently, the array factor A is responsible for the multilobe structure (fringes) of the receiving pattern and F gives the pattern's slowly varying envelope (Fig. 5.8).

Since we are interested in the resolving properties of the interferometer and since they depend chiefly on A, we may, insofar as radiation falling within the central portion of the main lobe of F is concerned, set the factor F equal to unity and thus assume that the receiving pattern of the interferometer is given by $|A|$ alone. Accordingly, from

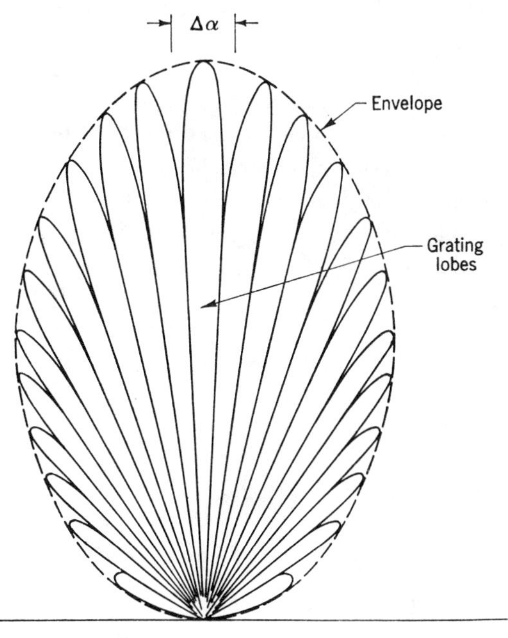

***Fig. 5.8** Polar plot of typical receiving pattern of two-element interferometer.*

Eq. (79) of Sec. 3.6 we see that the receiving pattern of the two-element interferometer is given by[1]

$$|A(\psi)| = 2\cos\left(\tfrac{1}{2}kl\cos\psi\right) \tag{153}$$

where ψ is the angle between the direction of the incident wave and the base line, i.e., the straight line joining the two antennas. In terms of the complementary angle $\alpha(= \pi/2 - \psi)$, i.e., the angle the direction of the source makes with the plane perpendicular to the base line (axial plane), the radiation pattern is

$$|A(\alpha)| = 2\cos\left(\tfrac{1}{2}kl\sin\alpha\right) \tag{154}$$

The power P fed to the receiver of the interferometer is proportional to $|A(\alpha)|^2$ and hence

$$P(\alpha) = 2P_0\cos^2\left(\tfrac{1}{2}kl\sin\alpha\right) = P_0[1 + \cos(kl\sin\alpha)] \tag{155}$$

where P_0 denotes the power fed to the receiver by a single antenna. As a point source of radiation sweeps across the sky, the angle α changes and P oscillates between the limits 0 and $2P_0$. This is strictly true for small α only. Actually, when α becomes large, the power fed to the receiver is no longer given by expression (155) alone, but by the product of expression (155) and $|F|^2$. The factor $|F|^2$ has the effect of tapering off the oscillations (Fig. 5.9). The nulls of the receiving pattern occur where

$$kl\sin\alpha = (2n + 1)\pi \qquad (n = 0, 1, 2, \ldots) \tag{156}$$

and the maxima occur where

$$kl\sin\alpha = 2n\pi \qquad (n = 0, 1, 2, \ldots) \tag{157}$$

For small values of α, i.e., for values of α such that $\sin\alpha = \alpha$, the width

[1] We obtain this expression from Eq. (79) of Chap. 3 by setting $n = 2$ and $\gamma = 0$. The fact that the receiver of the interferometer is located at the electrical center of the transmission line connecting the two antennas requires that $\gamma = 0$.

of each grating lobe is given by the simple relation

$$\Delta\alpha = \frac{2\pi}{kl} = \frac{\lambda}{l} \tag{158}$$

which shows that as the spacing l is increased the width of each grating lobe is decreased. It also shows that for a fixed spacing the width of each grating lobe is decreased as the wavelength λ to which the receiver is tuned is decreased.

In the derivation of formula (155) it was tacitly assumed that the incident radiation comes from a point source. We now shed this restriction and consider the more realistic case where the source has angular extent. In this case the received power is given by

$$P(\alpha_0) = \int [1 + \cos (kl \sin \alpha)] f(\alpha - \alpha_0) d\alpha \tag{159}$$

where $f(\alpha - \alpha_0)$ is the distribution across the incoherent source and α_0 is the angle that the mean direction of the source makes with the axial plane. If the width of the source is $2w$, the limits of integration are $\alpha = \alpha_0 - w$ and $\alpha = \alpha_0 + w$. We assume that α_0 and $2w$ are small, i.e., we assume that the source is narrow and near the axial plane. Expression (159) is a generalization of expression (155) and reduces to it when $f(\alpha - \alpha_0)$ is the Dirac delta function $\delta(\alpha - \alpha_0)$.

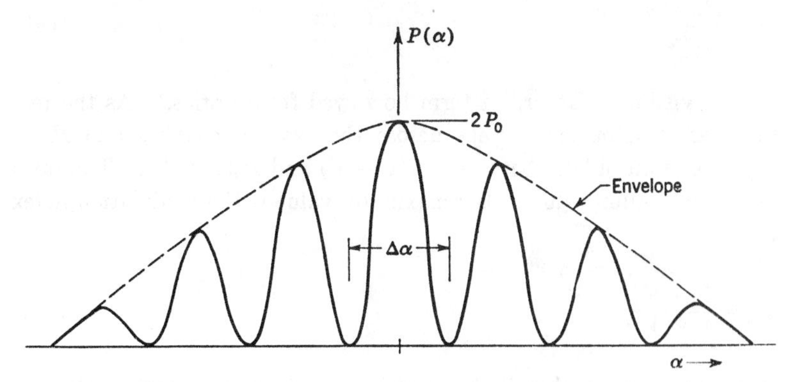

Fig. 5.9 ***Rectangular plot of receiving pattern of two-element interferometer for point source. The minima are zero. The maxima are tapered, by virtue of the fact that $|F|^2$ is not equal to one for all values of α. Actually, $|F|^2$ behaves in a manner indicated by the envelope.***

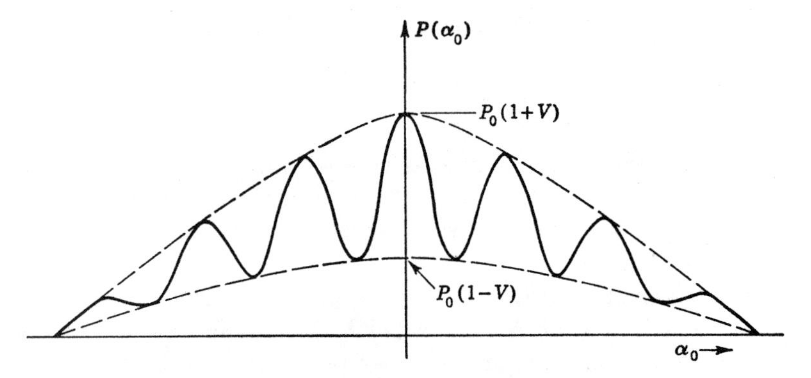

Fig. 5.10 Rectangular plot of receiving pattern for a narrow source having uniform distribution.

However, if $f(\alpha - \alpha_0)$ has a narrow rectangular shape, i.e., if $f(\alpha - \alpha_0) = P_0/2w$ for $\alpha_0 - w \leq \alpha \leq \alpha_0 + w$ and $f(\alpha - \alpha_0) = 0$ for all other values of α, expression (159) leads to

$$P(\alpha_0) = \int (1 + \cos kl\alpha) f(\alpha - \alpha_0) d\alpha = P_0(1 + V \cos kl\alpha_0) \tag{160}$$

where the quantity V defined by

$$V = \frac{\sin klw}{klw} \tag{161}$$

is the "visibility factor," a term borrowed from optics.[1] As the rectangular distribution sweeps across the sky, α_0 changes and $P(\alpha_0)$ oscillates sinusoidally between $P_0(1 - V)$ and $P_0(1 + V)$. The ratio of the minimum value to the maximum value is the modulation index M given by

$$M = \frac{1 - V}{1 + V} \tag{162}$$

From this we see that if the distribution function is rectangular the width of the source can be determined by measuring M and then computing w from Eqs. (161) and (162). See Fig. 5.10.

[1] See, for example, M. Born and E. Wolf, "Principles of Optics," pp. 264–267, Pergamon Press, New York, 1959.

More generally, if the source is narrow but otherwise arbitrary, it follows from Eq. (159) that for small values of α_0 the received power is given by

$$\begin{aligned} P(\alpha_0) &= \int [1 + \cos(kl \sin \alpha)] f(\alpha - \alpha_0) d\alpha \\ &= \int [1 + \cos(kl\alpha)] f(\alpha - \alpha_0) d\alpha \\ &= \int f(\alpha - \alpha_0) d\alpha + \int \cos(kl\alpha) f(\alpha - \alpha_0) d\alpha \end{aligned} \tag{163}$$

The first term on the right is P_0, the power fed to the receiver by one antenna; the second term on the right we denote by P_1. Accordingly, we write

$$P(\alpha_0) = P_0 + P_1 \tag{164}$$

where

$$P_1 = \int \cos(kl\alpha) f(\alpha - \alpha_0) d\alpha \tag{165}$$

If we let $u = \alpha - \alpha_0$ and note that $f(u) \equiv 0$ for $|u| > w$, then P_1 can be cast in a form that explicitly displays its amplitude and phase, viz.,

$$\begin{aligned} P_1 = \int \cos[kl(u + \alpha_0)] f(u) du &= \operatorname{Re} e^{ikl\alpha_0} \int_{-\infty}^{\infty} e^{iklu} f(u) du \\ &= \operatorname{Re} e^{ikl\alpha_0} Q(kl) e^{i\phi(kl)} = Q(kl) \cos[kl\alpha_0 + \phi(kl)] \end{aligned} \tag{166}$$

Here the amplitude $Q(kl)$ and the phase $\phi(kl)$ are defined by

$$Q(kl) e^{i\phi(kl)} = \int_{-\infty}^{\infty} e^{iklu} f(u) du \tag{167}$$

The inverse Fourier transform of Eq. (167) yields the relation

$$f(u) = \frac{1}{\pi} \int_0^{\infty} Q(kl) \cos[klu - \phi(kl)] d(kl) \tag{168}$$

which expresses $f(u)$ in terms of the amplitude and phase of the observed quantity P_1, viz., $Q(kl)$ and $\phi(kl)$. Relation (168) shows that it is possible, in principle, to find the distribution by measuring the amplitude and phase with different base lines. However, the measurement of phase sometimes presents difficulties. Unfortunately, it is not possible to determine uniquely the distribution from a knowledge of the amplitude alone, unless some information is available beforehand about the general shape of the distribution function.

As an example of a two-element radio interferometer with a horizontal base line we mention the one in Owens Valley, California, which is operated by the California Institute of Technology. Each element of the interferometer is a steerable parabolic reflector antenna, 90 feet in diameter, placed on the ground. It is used for the measurement of angular diameters at centimeter and decimeter wavelengths, and for positional work.[1]

A two-element interferometer having a vertical base line can be effected by placing a single horizontally beamed antenna on a cliff of height $l/2$ overlooking the sea. The surface of the sea acts as an image plane. Thus the elevated antenna and its image constitute a two-element interferometer.[2] The elevated antenna is horizontally polarized to take advantage of the fact that the surface of the sea approximates a perfect reflector most closely for horizontal polarization. The image antenna is out of time-phase with respect to the elevated antenna and hence the power received from a point source is given by

$$P(\alpha) = 2P_0[1 - \cos(kl \sin \alpha)] \tag{169}$$

where α is the angle the direction of the source makes with the axial plane, i.e., the surface of the sea, and P_0 is the power the elevated antenna would receive if it were not operating as an interferometer. In this case the nulls of the receiving pattern occur where

$$kl \sin \alpha = 2n\pi \qquad (n = 0, 1, 2, \ldots) \tag{170}$$

and the maxima occur where

$$kl \sin \alpha = (2n + 1)\pi \qquad (n = 0, 1, 2, \ldots) \tag{171}$$

[1] For details the reader is referred to J. G. Bolton, Radio Telescopes, chap. 11 in G. P. Kuiper and B. M. Middlehurst (eds.), "Telescopes," The University of Chicago Press, Chicago, 1960.

[2] An interferometer of this type is called a "sea interferometer," a "cliff interferometer," or a "Lloyd's mirror" after its optical analog.

From Eq. (169) we see that as the source rises above the horizon and cuts through the grating lobes of the interferometer, the received power increases from zero and oscillates in characteristic fashion. Then as the source rises above and out of the beam, the received power gradually tapers to zero, an effect which would have been displayed by Eq. (169) had it been multiplied by $|F|^2$.

5.8 Correlation Interferometer

The two-element interferometer discussed in the previous section behaves as though the incident radiation were monochromatic because the receiver of the interferometer is sharply tuned and accepts only a very narrow band of the incident radiation's broad spectrum. Since the energy residing outside this band is rejected and thus wasted, the sensitivity of the interferometer is limited by the bandwidth of the receiver. Increasing the bandwidth would increase the sensitivity but would also deteriorate the multilobe pattern and hence decrease the precision of the system. This means that in a phase-comparison type of interferometer the bandwidth is necessarily narrow and the sensitivity is limited by the bandwidth. In addition to this inherent limitation on the sensitivity there is a practical limitation on the resolving power. As the antennas are moved farther apart for the purpose of increasing the resolving power, it becomes more difficult to compare accurately the phases of the antenna outputs. The awkwardness of measuring the phases of two widely separated signals places a practical limitation on the antenna separation and this in turn places a limitation on the resolving power.

Because of these and other limitations, the two-element phase-comparison interferometer has been superseded in certain applications by more sophisticated systems. In this section we shall discuss one such system, namely, the correlation interferometer of Brown and Twiss.[1] But before we do this, let us discuss the concept of degree of coherence[2] upon which it is based.

[1] R. H. Brown and R. Q. Twiss, A New Type of Interferometer for Use in Radio Astronomy, *Phil. Mag.*, **45:** 663 (1954).

[2] F. Zernike, The Concept of Degree of Coherence and Its Application to Optical Problems, *Physica*, **5:** 785 (1938).

To measure the degree of coherence of the polychromatic radiation from an extended source we use two identical and similarly oriented antennas. These antennas receive the incident radiation and consequently develop at their respective output terminals the voltages $V_1(t)$ and $V_2(t)$, which for mathematical convenience are assumed to have the form of an analytical signal. The resulting voltages are fed into a receiver whose output is the time-average power given by

$$P = \langle [V_1(t) + V_2(t)][V_1^*(t) + V_2^*(t)] \rangle$$
$$= \langle V_1(t)V_1^*(t)\rangle + \langle V_2(t)V_2^*(t)\rangle + 2 \operatorname{Re} \langle V_1(t)V_2^*(t)\rangle \quad (172)$$

The first term on the right is the time-average power output of one antenna operating singly, and the second term is the time-average power of the other antenna operating singly. Hence, the third term is the only one that involves the mutual effects or mutual coherence of the incident radiation. Accordingly, as a quantitative measure of the mutual coherence of the incident radiation, we choose the complex quantity γ defined by

$$\gamma = \frac{\langle V_1(t)V_2^*(t)\rangle}{\sqrt{\langle V_1(t)V_1^*(t)\rangle\langle V_2(t)V_2^*(t)\rangle}} \quad (173)$$

and referred to as the complex degree of coherence. The modulus $|\gamma|$ of γ is known as the degree of coherence. By the Schwarz inequality it can be shown that

$$|\gamma| \leq 1 \quad (174)$$

When $|\gamma| = 0$ the incident radiation is incoherent; when $|\gamma| = 1$ the incident radiation is coherent; and when $0 < |\gamma| < 1$ the incident radiation is partially coherent. In terms of γ, expression (172) for the power output of the receiver becomes

$$P = \langle V_1(t)V_1^*(t)\rangle + \langle V_2(t)V_2^*(t)\rangle$$
$$+ 2\sqrt{\langle V_1(t)V_1^*(t)\rangle\langle V_2(t)V_2^*(t)\rangle}\, |\gamma| \cos(\arg\gamma) \quad (175)$$

where $\arg\gamma$ is the phase of γ, that is, $\gamma = |\gamma| \exp(i \arg\gamma)$. For simplicity we assume that the power outputs of the antennas when operated separately are equal; that is, we assume that

$$\langle V_1(t)V_1^*(t)\rangle = \langle V_2(t)V_2^*(t)\rangle = P_0 \tag{176}$$

With the aid of this assumption expression (175) reduces to the relation

$$P = 2P_0[1 + |\gamma| \cos(\arg \gamma)] \tag{177}$$

which clearly indicates that the degree of coherence $|\gamma|$ of the incident radiation is measured by the visibility. Expression (177) provides an operational definition of the degree of coherence.

To show how γ is related to data that specify the source and to the spacing of the antennas we proceed as follows. We choose a cartesian coordinate system with origin O in order that the antennas be located along the x axis at the points $x = \mp l/2$. For simplicity the source is assumed to be a line source lying along the ξ axis of a parallel cartesian system with origin O'. The distance between O and O' is R. See Fig. 5.11. We think of the source as being divided into elements of length $d\xi_1, d\xi_2, d\xi_3, \ldots$, and we denote the respective antenna output voltages due to the radiation from the mth element by the analytic signals $V_{m1}(t)$ and $V_{m2}(t)$. The respective antenna output voltages due to the radiation from the entire source are given by the sums

$$V_1(t) = \sum_m V_{ml}(t) \qquad V_2(t) = \sum_m V_{m2}(t) \tag{178}$$

We assume that each element of the source is an isotropic radiator. Consequently the radiation from the mth element produces the voltage

$$V_{m1}(t) = A_m\left(t - \frac{R_{m1}}{c}\right)\frac{e^{-i\omega(t-R_{m1}/c)}}{R_{m1}} \tag{179}$$

in one antenna and the voltage

$$V_{m2}(t) = A_m\left(t - \frac{R_{m2}}{c}\right)\frac{e^{-i\omega(t-R_{m2}/c)}}{R_{m2}} \tag{180}$$

in the other. Here R_{m1} and R_{m2} are the distances from the mth element to the antennas, c is the velocity of light, ω is the mean frequency

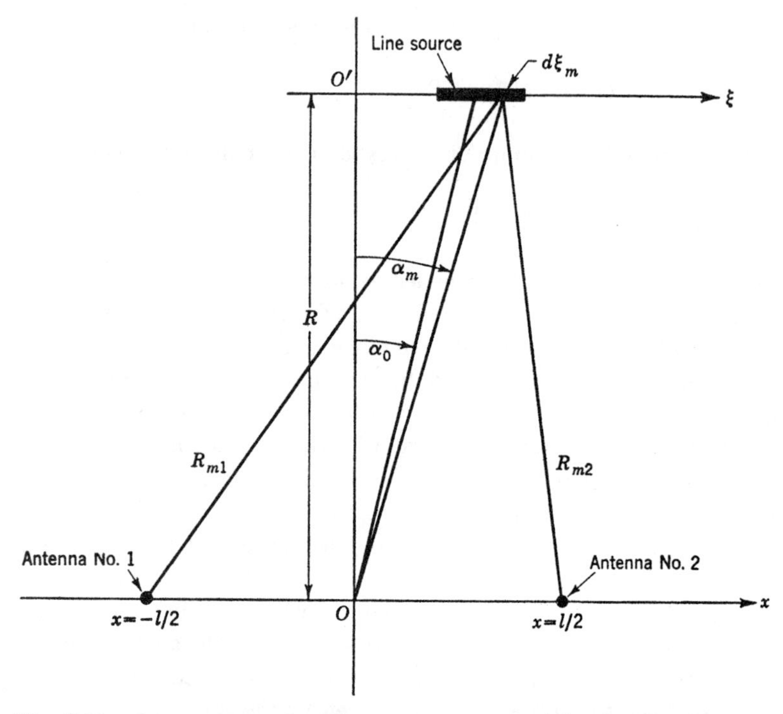

Fig. 5.11 Arrangement for the measurement of degree of coherence. Two identical and similarly oriented antennas are exposed to the polychromatic radiation from line source. R is distance from O to O'. The angle that the line connecting O with the center of the line source makes with the axial line OO' is α_0. The angle that the line from O to the element $d\xi_m$ makes with the line OO' is α_m. The distances from the element $d\xi_m$ to the antennas are R_{m1} and R_{m2}.

of the incident radiation, and A_m is the complex amplitude function.

It follows from expressions (178) that

$$\langle V_1(t)V_1^*(t)\rangle = \sum_m \langle V_{m1}(t)V_{m1}^*(t)\rangle + \sum_{m \neq n}\sum \langle V_{m1}(t)V_{n1}^*(t)\rangle \tag{181}$$

However, the isotropic radiators that make up the source are assumed to be statistically independent and to have a mean value of zero, i.e.,

$$\langle V_{m1}(t)V_{n1}^*(t)\rangle = \langle V_{m1}(t)\rangle\langle V_{n1}^*(t)\rangle = 0 \qquad \text{when } m \neq n \tag{182}$$

and consequently the cross-product terms of Eq. (181) vanish. Thus we get

$$\langle V_1(t)V_1^*(t)\rangle = \sum_m \langle V_{m1}(t)V_{m1}^*(t)\rangle \tag{183}$$

Similarly, we obtain

$$\langle V_2(t)V_2^*(t)\rangle = \sum_m \langle V_{m2}(t)V_{m2}^*(t)\rangle \tag{184}$$

and

$$\langle V_1(t)V_2^*(t)\rangle = \sum_m \langle V_{m1}(t)V_{m2}^*(t)\rangle \tag{185}$$

Substituting expressions (179) and (180) into Eqs. (183) and (184) respectively, and noting that A_m is stationary, we see that

$$\begin{aligned}\langle V_1(t)V_1^*(t)\rangle &= \sum_m \frac{1}{R_{m1}^2}\left\langle A_m\left(t - \frac{R_{m1}}{c}\right)A_m^*\left(t - \frac{R_{m1}}{c}\right)\right\rangle \\ &= \sum_m \frac{1}{R_{m1}^2}\langle A_m(t)A_m^*(t)\rangle\end{aligned} \tag{186}$$

and

$$\begin{aligned}\langle V_2(t)V_2^*(t)\rangle &= \sum_m \frac{1}{R_{m2}^2}\left\langle A_m\left(t - \frac{R_{m2}}{c}\right)A_m^*\left(t - \frac{R_{m2}}{c}\right)\right\rangle \\ &= \sum_m \frac{1}{R_{m2}^2}\langle A_m(t)A_m^*(t)\rangle\end{aligned} \tag{187}$$

Since R_{m1} and R_{m2} are approximately equal, these two expressions in this approximation are equal to each other and to P_0. That is, in agreement with assumption (176) we have

$$\langle V_1(t)V_1^*(t)\rangle = \langle V_2(t)V_2^*(t)\rangle = P_0 \tag{188}$$

Substituting expressions (179) and (180) into Eq. (185), we obtain

$$\langle V_1(t)V_2^*(t)\rangle = \sum_m \left\langle A_m\left(t - \frac{R_{m1}}{c}\right)A_m^*\left(t - \frac{R_{m2}}{c}\right)\right\rangle \frac{e^{ik(R_{m1}-R_{m2})}}{R_{m1}R_{m2}} \tag{189}$$

where $k = \omega/c$. Since R_{m1} and R_{m2} are approximately equal, and since A_m is stationary, Eq. (189) reduces to

$$\langle V_1(t)V_2^*(t)\rangle = \sum_m \langle A_m(t)A_m^*(t)\rangle \frac{e^{ik(R_{m1}-R_{m2})}}{R_{m1}R_{m2}} \tag{190}$$

To cast this expression into the form of an integral, we introduce the following geometric considerations. From Fig. 5.11 it is clear that

$$R_{m1}{}^2 = \left(\xi_m + \frac{l}{2}\right)^2 + R^2 \qquad R_{m2}{}^2 = \left(\xi_m - \frac{l}{2}\right)^2 + R^2 \tag{191}$$

Since $R \gg \left(\xi_m + \frac{l}{2}\right)$ and $R \gg \left(\xi_m - \frac{l}{2}\right)$, it follows from Eqs. (191) that

$$R_{m2} - R_{m1} = -\frac{\xi_m l}{R} \qquad \text{and} \qquad R_{m2}R_{m1} = R^2 \tag{192}$$

Using these approximations, we see that Eq. (190) becomes

$$\langle V_1(t)V_2^*(t)\rangle = \sum_m \frac{1}{R^2} \langle A_m(t)A_m^*(t)\rangle e^{ikl\xi_m/R} \tag{193}$$

Moreover, from Fig. 5.11 it is clear that $\tan \alpha_m = \xi_m/R$, but since α_m is small we have the simpler relation

$$\alpha_m = \frac{\xi_m}{R} \tag{194}$$

With the aid of relation (194) we may cast Eq. (193) in the following form:

$$\langle V_1(t)V_2^*(t)\rangle = \sum_m \frac{1}{R^2} \langle A_m(t)A_m^*(t)\rangle e^{ikl\alpha_m} \tag{195}$$

which suggests that the sum may be written as an integral. If we let

$$\frac{1}{R^2} \langle A_m(t)A_m^*(t)\rangle = f(\alpha_m - \alpha_0)d\alpha_m$$

then Eq. (195) in the limit becomes

$$\langle V_1(t) V_2^*(t) \rangle = \int f(\alpha - \alpha_0) e^{ikl\alpha}\, d\alpha \tag{196}$$

where α_0 is the angle that the line connecting O with the center of the source makes with the axial plane.

Substituting Eqs. (188) and (196) into expression (173), we find that the complex degree of coherence is related to the source distribution function $f(\alpha - \alpha_0)$ and to the antenna spacing kl by the relation

$$\gamma = \frac{1}{P_0} \int f(\alpha - \alpha_0) e^{ikl\alpha}\, d\alpha \tag{197}$$

Since

$$P_0 = \int f(\alpha - \alpha_0) d\alpha \tag{198}$$

we may also write γ in the homogeneous form

$$\gamma = \frac{\int f(\alpha - \alpha_0) e^{ikl\alpha}\, d\alpha}{\int f(\alpha - \alpha_0) d\alpha} \tag{199}$$

If in accord with the notation of the previous section we denote the amplitude and phase of the integral appearing in Eq. (196) by $Q(kl)$ and $\phi(kl)$ respectively, then we may write

$$\int f(\alpha - \alpha_0) e^{ikl\alpha}\, d\alpha = Q(kl) e^{ikl\alpha_0} e^{i\phi(kl)} \tag{200}$$

and from Eqs. (199) and (200) note that the degree of coherence is given by

$$|\gamma| = \frac{Q(kl)}{Q(O)} \tag{201}$$

and the phase of γ by

$$\arg \gamma = kl\alpha_0 + \phi(kl) \tag{202}$$

Substituting expressions (201) and (202) into Eq. (177), we obtain the expression

$$P = 2P_0 \left\{1 + \frac{Q(kl)}{Q(O)} \cos [kl\alpha_0 + \phi(kl)]\right\} \tag{203}$$

which places in evidence the equivalence of visibility and degree of coherence.

In the special case where $f(\alpha - \alpha_0)$ is a rectangular function of width $2w$, that is, $f(\alpha - \alpha_0) = P_0/2w$ for $\alpha_0 - w \leq \alpha \leq \alpha_0 + w$ and $f(\alpha - \alpha_0) = 0$ for all other values of α, we have

$$\int f(\alpha - \alpha_0)e^{ikl\alpha}\, d\alpha = \frac{P_0}{2w}\int_{\alpha_0 - w}^{\alpha_0 + w} e^{ikl\alpha}\, d\alpha = P_0 \frac{\sin klw}{klw} e^{ikl\alpha_0} \tag{204}$$

Hence for such a rectangular distribution the degree of coherence is given by

$$|\gamma| = \frac{Q(kl)}{Q(0)} = \frac{\sin klw}{klw} \tag{205}$$

and the phase of γ by

$$\arg \gamma = kl\alpha_0 \tag{206}$$

Thus we see that for a uniform source the magnitude γ is related by Eq. (205) to the width of the source and the phase $\arg \gamma$ is related by Eq. (206) to the angle between the axial plane and the line running from the origin to the center of the source.

The correlation coefficient $\rho(V_1, V_2)$ of $V_1(t)$ and $V_2(t)$ by definition is

$$\rho(V_1, V_2) = \frac{\langle (V_1 - \langle V_1 \rangle)(V_2^* - \langle V_2^* \rangle)\rangle}{\sigma(V_1)\sigma(V_2)} \tag{207}$$

where

$$\begin{aligned} \sigma^2(V_1) &= \langle (V_1 - \langle V_1 \rangle)(V_1^* - \langle V_1^* \rangle)\rangle \\ \sigma^2(V_2) &= \langle (V_2 - \langle V_2 \rangle)(V_2^* - \langle V_2^* \rangle)\rangle \end{aligned} \tag{208}$$

are the variances. Since $V_1(t)$ and $V_2(t)$ have zero mean value, i.e.,

$$\langle V_1 \rangle = 0 \qquad \langle V_2 \rangle = 0 \tag{209}$$

the expression for ρ reduces to

$$\rho(V_1,V_2) = \frac{\langle V_1 V_2^* \rangle}{\sqrt{\langle V_1 V_1^* \rangle \langle V_2 V_2^* \rangle}} \tag{210}$$

Comparing expressions (173) and (210), we see that the complex degree of coherence γ is equal to the correlation coefficient $\rho(V_1,V_2)$. Hence what is actually measured in the above arrangement is the amplitude and phase of the correlation coefficient $\rho(V_1,V_2)$.

Now let us suppose that the circuits are changed (to a Brown and Twiss system) so that we can measure the correlation coefficient of the square of the moduli $M_1(t)$ and $M_2(t)$ of $V_1(t)$ and $V_2(t)$ respectively. The correlation coefficient $\rho(M_1{}^2,M_2{}^2)$ by definition is

$$\rho(M_1{}^2,M_2{}^2) = \frac{\langle (M_1{}^2 - \langle M_1{}^2 \rangle)(M_2{}^2 - \langle M_2{}^2 \rangle) \rangle}{\sigma(M_1{}^2)\sigma(M_2{}^2)} \tag{211}$$

where

$$\begin{aligned} \sigma^2(M_1{}^2) &= \langle (M_1{}^2 - \langle M_1{}^2 \rangle)^2 \rangle \\ \sigma^2(M_2{}^2) &= \langle (M_2{}^2 - \langle M_2{}^2 \rangle)^2 \rangle \\ M_1{}^2 &= V_1 V_1^* \qquad M_2{}^2 = V_2 V_2^* \end{aligned} \tag{212}$$

Under the assumption that the receiver noise is negligible compared to the desired signal, it can be shown by statistical calculations[1] that

$$\rho(M_1{}^2,M_2{}^2) = \rho(V_1,V_2)\rho^*(V_1,V_2) \tag{213}$$

But

$$\rho(V_1,V_2) = \gamma$$

and hence

$$\rho(M_1{}^2,M_2{}^2) = \gamma\gamma^* = |\gamma|^2 \tag{214}$$

[1] E. N. Bramley, Diversity Effects in Spaced-aerial Reception of Ionospheric Waves, *Proc. Inst. Elec. Engrs.*, **98** (3): 9–25 (1951); also, J. A. Ratcliffe, Some Aspects of Diffraction Theory and their Application to the Ionosphere, *Rept. Prog. Phys.*, **19**: 188–267 (1956).

This means that the correlation coefficient of the squares of the moduli of the antenna voltages is equal to the square of the degree of coherence of the incident radiation. In the case where the source is a rectangular distribution of width $2w$, we see from Eqs. (205) and (214) that

$$\rho(M_1{}^2,M_2{}^2) = \left|\frac{\sin klw}{klw}\right|^2 \tag{215}$$

With the aid of this result w can easily be computed from a knowledge of the correlation coefficient $\rho(M_1{}^2,M_2{}^2)$.

Although $\rho(M_1{}^2,M_2{}^2)$ yields information about $|\gamma|$ only, and $\rho(V_1,V_2)$ yields information about $|\gamma|$ as well as $\arg \gamma$, the former is easier to measure, as no phase-preserving link between the antennas is required.

The correlation interferometer of Brown and Twiss may be defined as an interferometer that measures $\rho(M_1{}^2,M_2{}^2)$. It differs from a conventional interferometer, which measures $\rho(V_1,V_2)$. Since no radio-frequency phase-preserving link is necessary in the measurement of $\rho(M_1{}^2,M_2{}^2)$, the antennas can be separated greatly and thus high resolving powers can be realized.

Electromagnetic waves in a plasma 6

In recent years considerable attention has been focused on the theory of electromagnetic wave propagation in a plasma medium. In large measure this interest in the theory has been stimulated by its applicability to current problems in radio communications, radio astronomy, and controlled thermonuclear fusion. For example, the theory has been invoked to explain such phenomena as the propagation of radio waves in the ionosphere,[1] the propagation of cosmic radio waves in the solar atmosphere, in nebulae, and in interstellar and interplanetary space,[2] the reflection of radio waves from meteor trails[3] and from the envelope of ionized gas that surrounds a spacecraft as it penetrates

[1] K. G. Budden, "Radio Waves in the Ionosphere," Cambridge University Press, New York, 1961; also, J. A. Ratcliffe, "The Magneto-ionic Theory," Cambridge University Press, New York, 1961.

[2] V. L. Ginzburg, "Propagation of Electromagnetic Waves in Plasma," Gordon and Breach, Science Publishers, Inc., New York, 1961; also, I. S. Shklovsky, "Cosmic Radio Waves," Harvard University Press, Cambridge, Mass., 1960.

[3] N. Herlofson, Plasma Resonance in Ionospheric Irregularities, *Arkiv Fysik,* **3**: 247 (1951); also, J. L. Heritage, S. Weisbrod, and W. J. Fay, "Experimental Studies of Meteor Echoes at 200 Megacycles in Electromagnetic Wave Propagation," in M. Desirant and J. L. Michiels (eds.), Academic Press Inc., New York, 1960.

the atmosphere,[1] and the propagation of microwaves in laboratory plasmas.[2]

In these applications the medium through which the electromagnetic wave must travel is formally the same: it is a plasma, or more descriptively, a macroscopically neutral ionized gas consisting principally of free electrons, free ions, and neutral atoms or molecules. This means that from one application to another the nature of the problem does not change essentially, despite the large variations the medium may undergo in, say, its degree of ionization and its temperature. However, in the presence of a beam of charged particles interacting with the plasma, an electromagnetic wave does acquire characteristics which differ qualitatively from those in a beam-free plasma. One such characteristic is, for example, wave amplification by beam-generated plasma instabilities.[3] Accordingly, phenomena of this kind have to be treated separately and for this reason are excluded from the present discussion.

In this chapter we shall analyze the problem of electromagnetic wave propagation in a plasma medium by calculating the constitutive parameters of the plasma and then treating the problem as a conventional problem in the theory of electromagnetic wave propagation in a continuous medium.

6.1 Alternative Descriptions of Continuous Media

We recall from electromagnetic theory that for a continuous medium at rest Maxwell's equations can be written in the following elementary

[1] *Proc. Symp. Plasma Sheath,* vol. 1, U.S. Air Force, Cambridge Research Center, December, 1959.

[2] V. E. Goland, Microwave Plasma Diagnostic Techniques, *J. Tech. Phys., U.S.S.R.,* **30:** 1265 (1960).

[3] R. A. Demirkhanov, A. K. Gevorkov, and A. F. Popov, The Interaction of a Beam of Charged Particles with a Plasma, *Proc. Fourth Intern. Conf. on Ionization Phen. in Gases,* vol. 2, p. 665, North Holland Publishing Company, Amsterdam, August, 1959.

form,[1]

$$\frac{1}{\mu_0}\nabla \times \mathbf{B} = \mathbf{J}_t + \epsilon_0 \frac{\partial}{\partial t}\mathbf{E} \tag{1}$$

$$\nabla \times \mathbf{E} = -\frac{\partial}{\partial t}\mathbf{B} \tag{2}$$

$$\epsilon_0 \nabla \cdot \mathbf{E} = \rho_t \tag{3}$$

$$\nabla \cdot \mathbf{B} = 0 \tag{4}$$

which describes the macroscopic electromagnetic field in the medium by the two vector fields **E** and **B** and characterizes the medium by the total macroscopic charge density ρ_t and total macroscopic current density $\mathbf{J}_t$. The constants μ_0 and ϵ_0 denote respectively the permeability and dielectric constant of the vacuum.

The total charge density ρ_t consists of the free charge density ρ and bound charge density ρ_b; similarly the total current density $\mathbf{J}_t$ consists of the free current density **J** and bound current density $\mathbf{J}_b$, that is,

$$\rho_t = \rho + \rho_b \tag{5}$$

$$\mathbf{J}_t = \mathbf{J} + \mathbf{J}_b \tag{6}$$

The free charge is that part of the total charge which exists independently of the field. On the other hand, the bound charge is an attribute of the multipoles that are induced in the medium by the electromagnetic field. Indeed, ρ_b and $\mathbf{J}_b$ are given by the series[2]

$$\rho_b = -\nabla \cdot \mathbf{P} + \tfrac{1}{2}\nabla\nabla : \mathbf{Q} + \cdots \tag{7}$$

$$\mathbf{J}_b = \frac{\partial}{\partial t}\mathbf{P} - \frac{1}{2}\frac{\partial}{\partial t}\nabla \cdot \mathbf{Q} + \nabla \times \mathbf{M} + \cdots \tag{8}$$

[1] See, for example, R. W. P. King, "Electromagnetic Engineering," McGraw-Hill Book Company, New York, 1945; also, L. Rosenfeld, "Theory of Electrons," North Holland Publishing Company, Amsterdam, 1951.

[2] We keep only the leading terms. When the series are terminated at a certain degree of approximation, the number of electric multipoles exceeds the number of magnetic multipoles by one. In compliance with this rule we have kept two electric multipoles **P** and **Q** and one magnetic multipole **M**.

where **P**, **M**, **Q** denote respectively the volume densities of the electric dipoles, magnetic dipoles, and electric quadrupoles that are produced by the action of the electromagnetic field on the neutral molecules of the medium. In other words, **P**, **M**, **Q** are functionals of **E** and **B**.

In view of the series (7) and (8), Maxwell's equations (1), (2), (3), (4) become

$$\frac{1}{\mu_0}\nabla\times\mathbf{B} = \mathbf{J} + \epsilon_0\frac{\partial}{\partial t}\mathbf{E} + \frac{\partial}{\partial t}\mathbf{P} - \frac{1}{2}\frac{\partial}{\partial t}\nabla\cdot\mathbf{Q} + \nabla\times\mathbf{M} + \cdots \tag{9}$$

$$\nabla\times\mathbf{E} = -\frac{\partial}{\partial t}\mathbf{B} \tag{10}$$

$$\epsilon_0\nabla\cdot\mathbf{E} = \rho - \nabla\cdot\mathbf{P} + \tfrac{1}{2}\nabla\nabla:\mathbf{Q} + \cdots \tag{11}$$

$$\nabla\cdot\mathbf{B} = 0 \tag{12}$$

If we define the electric displacement **D** by

$$\nabla\cdot\mathbf{D} = \rho \tag{13}$$

then on comparing this relation with Eq. (11) we see that this definition leads to

$$\mathbf{D} = \epsilon_0\mathbf{E} + \mathbf{P} - \tfrac{1}{2}\nabla\cdot\mathbf{Q} + \cdots \tag{14}$$

Moreover, if we define the vector **H** by

$$\mathbf{H} = \frac{1}{\mu_0}\mathbf{B} - \mathbf{M} \tag{15}$$

then Eq. (9) leads to

$$\nabla\times\mathbf{H} = \mathbf{J} + \frac{\partial}{\partial t}\mathbf{D} \tag{16}$$

Hence, when **D** is defined by Eq. (13) and **H** by Eq. (15), the Maxwell equations (9), (10), (11), (12) assume their conventional form:

$$\nabla\times\mathbf{H} = \mathbf{J} + \frac{\partial}{\partial t}\mathbf{D} \tag{17}$$

$$\nabla\times\mathbf{E} = -\frac{\partial}{\partial t}\mathbf{B} \tag{18}$$

$$\nabla\cdot\mathbf{D} = \rho \tag{19}$$

$$\nabla\cdot\mathbf{B} = 0 \tag{20}$$

To apply these considerations to the case of an electromagnetic wave passing through a plasma medium, we note that the wave, in principle, interacts with all three components of the plasma, viz., the free electrons, the free ions, and the neutral molecules. However, the interaction of the wave with the neutral particles is so feeble in comparison to the interaction between the wave and the charged particles that it can be neglected. This means that **P**, **M**, **Q**, which constitute a measure of the interaction between the wave and the neutral particles, can be set equal to zero. Moreover, since the ions are much more massive than the electrons, the velocity imparted to the ions by the wave is negligibly small compared to the velocity given to the electrons. That is, when an electromagnetic wave passes through a sufficiently ionized plasma only the free electrons of the plasma influence appreciably the transmission of the wave.

The interaction between the wave and the electrons is introduced into Maxwell's equations through the current density term **J**. As will be shown subsequently (see Eq. 43), the electronic current density **J** produced in the plasma by the wave is related in the steady state to the electric vector **E** of the wave by a linear relation of the form

$$\mathbf{J} = a\mathbf{E} + i\omega b\mathbf{E} \qquad (a,\, b = \text{positive real}) \tag{21}$$

unless **E** exceeds a value at which nonlinearities come into play. It therefore follows that when an electromagnetic wave whose electric vector **E** lies within the bound of linearity passes through a sufficiently ionized plasma, the Maxwell equations for the phenomenon in the steady state become

$$\nabla \times \mathbf{H} = a\mathbf{E} + i\omega b\mathbf{E} - i\omega\epsilon_0\mathbf{E} \tag{22}$$

$$\nabla \times \mathbf{E} = i\omega\mu_0\mathbf{H} \tag{23}$$

Let us write Eq. (22) in the form

$$\nabla \times \mathbf{H} = (a + i\omega b)\mathbf{E} - i\omega\epsilon_0\mathbf{E} \tag{24}$$

where $(a + i\omega b)\mathbf{E}$ appears as a conduction current and $-i\omega\epsilon_0\mathbf{E}$ as a vacuum displacement current. This form suggests that we think of the complex factor $(a + i\omega b)$ as a complex conductivity given by

$$\sigma_c = \sigma_r + i\sigma_i = a + i\omega b \tag{25}$$

and thus describe the plasma as a conductor having a permeability μ_0, a dielectric constant ϵ_0, and a complex conductivity σ_c. However, we shall not use this mode of description here. Instead, we shall interpret the term $i\omega b\mathbf{E}$ of Eq. (22) as a polarization current and thus consider the plasma as a lossy dielectric.

To do this, we recall that for a lossy dielectric by definition we have

$$\nabla \times \mathbf{H} = \sigma\mathbf{E} - i\omega\mathbf{P} - i\omega\epsilon_0\mathbf{E} \tag{26}$$

where σ is the conductivity of the dielectric and $\mathbf{P}$ is the polarization of the neutral molecules of the dielectric. Also for a dielectric we have

$$\mathbf{P} = \chi_e\mathbf{E} \tag{27}$$

where χ_e, the electric susceptibility of the dielectric, is always positive. Since the relation

$$\mathbf{D} = \epsilon_0\mathbf{E} + \mathbf{P} = \epsilon\mathbf{E} \tag{28}$$

defines the dielectric constant ϵ of the dielectric, it follows that the dielectric constant of the dielectric is given by

$$\epsilon = \epsilon_0 + \chi_e \tag{29}$$

Clearly, for a true dielectric ϵ is always greater than ϵ_0 because $\chi_e \geq 0$.

If we are to describe the plasma as a lossy dielectric, we must identify Eq. (22) with Eq. (26), setting $a\mathbf{E} = \sigma\mathbf{E}$ and $i\omega b\mathbf{E} = -i\omega\mathbf{P} = -i\omega\chi_e\mathbf{E}$. This means that the conductivity σ of the dielectric must equal a and its electric susceptibility χ_e must equal $-b$, that is, $\sigma = a$ and $\chi_e = -b$. Since b is positive, χ_e must be negative. Thus, if the effect of the motion of the electrons is to be accounted for by a conductivity and a polarization, then we must think of the plasma as a lossy dielectric whose electric susceptibility is negative. The constitutive parameters of the dielectric are then given by

$$\sigma = a \qquad \mu = \mu_0 \qquad \epsilon = \epsilon_0 - b \tag{30}$$

Here we note that in contrast to an actual dielectric ϵ is less than ϵ_0.

Also we may combine the conductivity with the dielectric constant and thus obtain a complex dielectric constant ϵ_c. If this is done, the plasma is described by the constitutive parameters

$$\mu = \mu_0 \qquad \epsilon_c = -\frac{a}{i\omega} + \epsilon_0 - b \tag{31}$$

6.2 Constitutive Parameters of a Plasma

When a high-frequency electromagnetic wave passes through a plasma, only the interaction between the wave and the free electrons need be considered. Therefore, from a statistical point of view the macroscopic state of the plasma can be described in terms of a single distribution function $f(\mathbf{r},\mathbf{w},t)$, which determines the probable number of electrons that at the time t lie within the spatial volume $dx\,dy\,dz$ centered at $\mathbf{r}$ and have velocities within the intervals dw_x, dw_y, dw_z centered at $\mathbf{w}$.

This function of the position vector $\mathbf{r}$, the velocity vector $\mathbf{w}$, and the time t must satisfy the Boltzmann (or kinetic) equation

$$\frac{df}{dt} \equiv \frac{\partial f}{\partial t} + \mathbf{w} \cdot \nabla f + \left(\frac{d}{dt}\mathbf{w}\right) \cdot \nabla_w f = C \tag{32}$$

where $\nabla_w f$ is the gradient of f in velocity space, ∇f is the gradient of f in coordinate space, and C is the temporal rate of change in f caused by collisions. The acceleration $d\mathbf{w}/dt$ is related to $\mathbf{E}$ and $\mathbf{B}$ of the wave in accord with the Lorentz force equation

$$m\frac{d}{dt}\mathbf{w} = q(\mathbf{E} + \mathbf{w} \times \mathbf{B}) \tag{33}$$

where q and m denote respectively the charge and mass of the electron. Substituting expression (33) into Eq. (32), we obtain

$$\frac{\partial f}{\partial t} + \mathbf{w} \cdot \nabla f + \frac{q}{m}(\mathbf{E} + \mathbf{w} \times \mathbf{B}) \cdot \nabla_w f = C \tag{34}$$

which shows explicitly that the driving force is the macroscopic electromagnetic field $\mathbf{E}$, $\mathbf{B}$. Multiplying this equation by $m\mathbf{w}$ and integrating over all velocities, we obtain[1] the macroscopic equation of motion

$$nm\left(\frac{\partial}{\partial t}\mathbf{v} + \mathbf{v}\cdot\nabla\mathbf{v}\right) = nq(\mathbf{E} + \mathbf{v}\times\mathbf{B}) - \nabla\cdot\mathbf{S} + \mathbf{G} \tag{35}$$

In this equation the particle density $n(\mathbf{r},t)$ and the macroscopic velocity $\mathbf{v}(\mathbf{r},t)$ are defined respectively by

$$n(\mathbf{r},t) = \iiint_{-\infty}^{\infty} f(\mathbf{r},\mathbf{w},t)\,dw_x dw_y dw_z \tag{36}$$

$$\mathbf{v}(\mathbf{r},t) = \frac{1}{n}\iiint_{-\infty}^{\infty} \mathbf{w}f(\mathbf{r},\mathbf{w},t)\,dw_x dw_y dw_z \tag{37}$$

The dyadic $\mathbf{S}$ is the stress, defined by

$$\mathbf{S} = m\iiint_{-\infty}^{\infty} (\mathbf{w} - \mathbf{v})(\mathbf{w} - \mathbf{v})f(\mathbf{r},\mathbf{w},t)\,dw_x dw_y dw_z \tag{38}$$

and the vector $\mathbf{G}$ is the net gain of momentum due to collisions.

In the present case all the nonlinear terms as well as the $\mathbf{v}\times\mathbf{B}$ term are dropped from Eq. (35), and thus the equation of motion is reduced, in the steady state, to the following simple form:

$$-i\omega nm\mathbf{v} = nq\mathbf{E} + \mathbf{G} \tag{39}$$

Moreover, since $\mathbf{G}$ is the net gain in momentum per unit volume per unit time, we may write

$$\mathbf{G} = -nm\mathbf{v}\omega_{eff} \tag{40}$$

where the proportionality constant ω_{eff} is the collision frequency and measures the number of effective collisions an electron makes per unit time. Furthermore, the density of electronic current $\mathbf{J}$ and the plasma

[1] See, for example, L. Spitzer, Jr., "Physics of Fully Ionized Gases," Interscience Publishers, Inc., New York, 1956.

frequency ω_p are defined by

$$\mathbf{J} = nq\mathbf{v} \tag{41}$$

and by

$$\omega_p{}^2 = \frac{nq^2}{m\epsilon_0} \tag{42}$$

Hence, from the equation of motion (39) and the expressions (40), (41), and (42) we find that the electronic current density **J** is related to **E** as follows:

$$\mathbf{J} = \frac{\epsilon_0\omega_p{}^2}{-i\omega + \omega_{eff}}\mathbf{E} = \frac{\epsilon_0\omega_{eff}\omega_p{}^2}{\omega^2 + \omega_{eff}{}^2}\mathbf{E} + i\omega\frac{\epsilon_0\omega_p{}^2}{\omega^2 + \omega_{eff}{}^2}\mathbf{E} \tag{43}$$

Comparing expression (43) with Eq. (21) of the previous section, we determine the coefficients a and b; and then by using relations (30) of the previous section, we find the constitutive parameters of the plasma. Accordingly, if we think of the plasma as a lossy dielectric, its conductivity is given by

$$\sigma = \frac{\epsilon_0\omega_{eff}\omega_p{}^2}{\omega^2 + \omega_{eff}{}^2} \tag{44}$$

its dielectric constant by

$$\epsilon = \epsilon_0\left(1 - \frac{\omega_p{}^2}{\omega^2 + \omega_{eff}{}^2}\right) \tag{45}$$

and its permeability by

$$\mu = \mu_0 \tag{46}$$

The elementary derivation of the constitutive parameters given above makes use of the collision frequency merely as an unknown parameter, without providing any information about its value. To evaluate ω_{eff}, the microprocesses which the plasma particles undergo must be taken into account explicitly. This has been done elsewhere by kinetic theory and the results show that ω_{eff} is not constant at all. Nevertheless, expressions (44) and (45) with ω_{eff} taken to be constant adequately describe the plasma for our present purposes.

6.3 Energy Density in Dispersive Media

Using Maxwell's equations for a lossless medium, we can write

$$\nabla \cdot \mathbf{S}(t) = -\mathbf{E}(t) \cdot \frac{\partial}{\partial t} \mathbf{D}(t) - \mathbf{H}(t) \cdot \frac{\partial}{\partial t} \mathbf{B}(t) \tag{47}$$

where

$$\mathbf{S}(t) = \mathbf{E}(t) \times \mathbf{H}(t) \tag{48}$$

is the Poynting vector. The quantity $\nabla \cdot \mathbf{S}(t)$ represents the rate of change of the electromagnetic energy density $w(t)$, that is,

$$\nabla \cdot \mathbf{S}(t) = -\frac{\partial}{\partial t} w(t) \tag{49}$$

From Eqs. (47) and (49) we see that

$$\frac{\partial w}{\partial t} = \mathbf{E}(t) \cdot \frac{\partial}{\partial t} \mathbf{D}(t) + \mathbf{H}(t) \cdot \frac{\partial}{\partial t} \mathbf{B}(t) \tag{50}$$

For a simple, nondispersive, lossless dielectric ϵ is a real constant and μ is equal to μ_0; hence

$$\mathbf{D}(t) = \epsilon \mathbf{E}(t) \qquad \mathbf{B}(t) = \mu_0 \mathbf{H}(t) \tag{51}$$

and relation (50) reduces to

$$\frac{\partial}{\partial t} w(t) = \frac{\partial}{\partial t} [\tfrac{1}{2}\epsilon \mathbf{E}(t) \cdot \mathbf{E}(t) + \tfrac{1}{2}\mu_0 \mathbf{H}(t) \cdot \mathbf{H}(t)] \tag{52}$$

which shows that the electromagnetic energy density for a simple, nondispersive, lossless dielectric is given by

$$w(t) = \tfrac{1}{2}\epsilon\mathbf{E}(t) \cdot \mathbf{E}(t) + \tfrac{1}{2}\mu_0\mathbf{H}(t) \cdot \mathbf{H}(t) \tag{53}$$

The first term on the right is the electric energy density w_e and the second term is the magnetic energy density w_m:

$$w_e(t) = \tfrac{1}{2}\epsilon\mathbf{E}(t) \cdot \mathbf{E}(t) \tag{54}$$

$$w_m(t) = \tfrac{1}{2}\mu_0\mathbf{H}(t) \cdot \mathbf{H}(t) \tag{55}$$

In the case of harmonic time dependence, where $\mathbf{E}(t) = \text{Re}\,\{\mathbf{E}e^{-i\omega t}\}$ and $\mathbf{H}(t) = \text{Re}\,\{\mathbf{H}e^{-i\omega t}\}$, the time-average energy densities may be written in terms of the phasors **E**, **H** as follows:

$$\bar{w}_m = \tfrac{1}{4}\mu_0\mathbf{H} \cdot \mathbf{H}^* \tag{56}$$

$$\bar{w}_e = \tfrac{1}{4}\epsilon\mathbf{E} \cdot \mathbf{E}^* \tag{57}$$

To define the electric and magnetic energy densities of an electromagnetic wave in a plasma, we must assume that the plasma is lossless, because it is only for a lossless medium that electromagnetic energy can be rationally defined as a thermodynamic quantity. For this reason we must limit our consideration to situations where the collision frequency ω_{eff} is so small that we may set it equal to zero. In keeping with this restriction, we consider a plasma whose collision frequency is zero and note that its constitutive parameters are

$$\epsilon = \epsilon_0\left(1 - \frac{\omega_p{}^2}{\omega^2}\right) \qquad \mu = \mu_0 \qquad \sigma = 0 \tag{58}$$

as can be seen by setting ω_{eff} equal to zero in Eqs. (44) and (45). Since μ is a constant, the magnetic energy density can be evaluated by means of relation (54) or (56). However, since ϵ is a function of frequency, the medium is dispersive and relations (55) and (57) no longer can be used to evaluate the electric energy density. For example, if we use relation (57) we obtain the expression

$$\bar{w}_e = \tfrac{1}{4}\epsilon_0\left(1 - \frac{\omega_p{}^2}{\omega^2}\right)\mathbf{E} \cdot \mathbf{E}^* \tag{59}$$

which predicts that $\bar{w}_e < 0$ when $\omega < \omega_p$, in contradiction to the fact that $\bar{w}_e$ must always be positive-definite.

Since the plasma is dispersive we cannot compute the electric energy density on a monochromatic basis. The reason for this is that since $\partial w_e/\partial t = \mathbf{E}(t) \cdot \partial \mathbf{D}(t)/\partial t$, the expression for the electric energy density, viz., $w_e(t) = \int \mathbf{E}(t) \cdot \partial \mathbf{D}(t)/\partial t \, dt + C$, contains the integration constant C, whose value depends on how the field is established. To determine C, we assume that the wave is quasi-monochromatic; then for $t \to -\infty$ we have $\mathbf{E}(-\infty) = 0$, $w_e(-\infty) = 0$, and hence $C = 0$. That is, for a quasi-monochromatic wave that starts in the remote past from value zero and builds up gradually, the integration constant is zero and $w_e(t)$ is fully determined.

A high-frequency wave whose amplitude is slowly modulated is a simple type of wave that builds up gradually in time and thus serves well in calculating electric energy density. Accordingly, we assume that the time dependence of the electric vector in the lossless plasma has the form

$$\mathbf{E}(t) = \tfrac{1}{2}\mathbf{E}_0[\cos(\omega + \Delta\omega)t - \cos(\omega - \Delta\omega)t] = -\mathbf{E}_0 \sin \Delta\omega t \sin \omega t \tag{60}$$

where $\mathbf{E}_0$ is a constant vector and $\Delta\omega$ is small compared to ω. Since $\mathbf{D} = \epsilon(\omega)\mathbf{E}$, the resulting displacement vector is

$$\mathbf{D}(t) = \tfrac{1}{2}\mathbf{E}_0[\epsilon(\omega + \Delta\omega)\cos(\omega + \Delta\omega)t - \epsilon(\omega - \Delta\omega)\cos(\omega - \Delta\omega)t] \tag{61}$$

and the resulting displacement current is

$$\frac{\partial}{\partial t}\mathbf{D}(t) = -\tfrac{1}{2}\mathbf{E}_0[(\omega + \Delta\omega)\epsilon(\omega + \Delta\omega)\sin(\omega + \Delta\omega)t - (\omega - \Delta\omega)\epsilon(\omega - \Delta\omega)\sin(\omega - \Delta\omega)t] \tag{62}$$

Expanding $(\omega + \Delta\omega)\epsilon(\omega + \Delta\omega)$ and $(\omega - \Delta\omega)\epsilon(\omega - \Delta\omega)$ in a Taylor series and retaining only the first two terms, we get the approximate expressions

$$(\omega + \Delta\omega)\epsilon(\omega + \Delta\omega) = \omega\epsilon + \Delta\omega \frac{\partial}{\partial \omega}(\epsilon\omega) + \cdots \tag{63}$$

$$(\omega - \Delta\omega)\epsilon(\omega - \Delta\omega) = \omega\epsilon - \Delta\omega \frac{\partial}{\partial \omega}(\epsilon\omega) + \cdots \tag{64}$$

which when substituted into expression (62) lead to the following expression for the displacement current:

$$\frac{\partial}{\partial t}\mathbf{D}(t) = -\mathbf{E}_0\left[\omega\epsilon \sin \Delta\omega t \cos \omega t + \Delta\omega \frac{\partial}{\partial\omega}(\omega\epsilon) \cos \Delta\omega t \sin \omega t\right] \tag{65}$$

We see from Eq. (50) that the rate of change of the electric energy density is

$$\frac{\partial}{\partial t} w_e = \mathbf{E}(t) \cdot \frac{\partial}{\partial t}\mathbf{D}(t) \tag{66}$$

and hence the energy gained during the time interval $t_1 - t_0$ is given by

$$w_e(t_1) - w_e(t_0) = \int_{t_0}^{t_1} \mathbf{E}(t) \cdot \frac{\partial}{\partial t}\mathbf{D}(t)dt \tag{67}$$

From expression (60) it is evident that $\mathbf{E}(t)$ is zero when $t = 0$ and has the form of a high-frequency carrier $\sin \omega t$ whose modulation envelope $\sin \Delta\omega t$ increases slowly with time. The time required for $\mathbf{E}(t)$ to build up from zero to its maximum value is $\Delta\omega t = \pi/2$ or $t = \pi/2\Delta\omega$. The energy gained during the time interval $t_0 = 0$ to $t_1 = \pi/2\Delta\omega$ is given by

$$w_e = \int_0^{\pi/2\Delta\omega} \mathbf{E}(t) \cdot \frac{\partial}{\partial t}\mathbf{D}(t)dt \tag{68}$$

Substituting expressions (60) and (65) into Eq. (68), we get

$$\begin{aligned} w_e = {} & \mathbf{E}_0 \cdot \mathbf{E}_0\omega\epsilon \int_0^{\pi/2\Delta\omega} \sin^2 \Delta\omega t \sin \omega t \cos \omega t \, dt \\ & + \mathbf{E}_0 \cdot \mathbf{E}_0 \, \Delta\omega \frac{\partial}{\partial\omega}(\omega\epsilon) \int_0^{\pi/2\Delta\omega} \sin^2 \omega t \sin \Delta\omega t \cos \Delta\omega t \, dt \end{aligned} \tag{69}$$

The first integral on the right is negligibly small compared to the second. In the second integral we may replace $\sin^2 \omega t$ by ½ and thus approximate the integral by

$$\tfrac{1}{2} \int_0^{\pi/2\Delta\omega} \sin \Delta\omega t \cos \Delta\omega t \, dt = \frac{1}{4\Delta\omega} \tag{70}$$

It follows that the time-average electric energy density is given by

$$\bar{w}_e = \tfrac{1}{4}\mathbf{E}_0 \cdot \mathbf{E}_0 \frac{\partial}{\partial\omega}(\omega\epsilon) \tag{71}$$

If instead of the form (60) for $\mathbf{E}(t)$ we take $\mathbf{E}(t) = \text{Re}\,\{\mathbf{E}_0(t)e^{-i\omega t}\}$, where $\mathbf{E}_0(t)$ is a slowly varying function, we would get again

$$\bar{w}_e = \frac{1}{4}\frac{\partial}{\partial\omega}(\omega\epsilon)\mathbf{E}_0 \cdot \mathbf{E}_0^* \tag{72}$$

Since $\epsilon = \epsilon_0(1 - \omega_p{}^2/\omega^2)$, expression (72) leads to

$$\bar{w}_e = \tfrac{1}{4}\epsilon_0\mathbf{E}_0 \cdot \mathbf{E}_0^* + \tfrac{1}{4}\epsilon_0 \frac{\omega_p{}^2}{\omega^2}\mathbf{E}_0 \cdot \mathbf{E}_0^* \tag{73}$$

which shows that $\bar{w}_e$ is the sum of two terms, the first representing the energy in the vacuum and the second representing the kinetic energy of the electrons.[1] To demonstrate that the second term does equal the time-average kinetic energy of the electrons, we recall from Eq. (39) of the previous section that for a lossless plasma

$$-i\omega nm\mathbf{v} = nq\mathbf{E} \tag{74}$$

The time-average kinetic energy density is, therefore, given by

$$\bar{K} = \tfrac{1}{4}nm\mathbf{v} \cdot \mathbf{v}^* = \frac{1}{4}\frac{nq^2}{\omega^2 m}\mathbf{E} \cdot \mathbf{E}^* \tag{75}$$

Using definition (42) of the plasma frequency, we get

$$\bar{K} = \tfrac{1}{4}\epsilon_0 \frac{\omega_p{}^2}{\omega^2}\mathbf{E} \cdot \mathbf{E}^* \tag{76}$$

which is identical with the second term of expression (73).

Thus we see that for a lossless plasma the time-average electro-

[1] Formally, this result can also be obtained from the energy theorem of Chap. 1; see F. Borgnis, Zur elektromagnetischen Energiedichte in Medien mit Dispersion, *Z. Physik,* **159:** 1–6 (1960).

magnetic energy density is given by[1]

$$\bar{w} = \tfrac{1}{4}\mu_0 \mathbf{H} \cdot \mathbf{H}^* + \frac{1}{4} \frac{\partial}{\partial \omega} (\omega \epsilon) \mathbf{E} \cdot \mathbf{E}^* \tag{77}$$

6.4 Propagation of Transverse Waves in Homogeneous Isotropic Plasma

To determine the propagation properties of transverse electromagnetic waves in a homogeneous isotropic plasma, we consider a linearly polarized plane transverse wave whose electric vector $E(t)$ has the form

$$\mathbf{E}(t) = \mathrm{Re}\,\{\mathbf{E}_0(t) e^{ikz} e^{-i\omega t}\} \tag{78}$$

where $\mathbf{E}_0(t)$ is a slowly varying function of time, ω is the real mean angular frequency, and k is the propagation constant or mean wave number, which may be complex. In a medium whose constitutive parameters are ϵ, μ_0, σ, the electric vector must satisfy

$$\nabla \times \nabla \times \mathbf{E} + \mu_0 \sigma \frac{\partial}{\partial t} \mathbf{E} + \mu_0 \epsilon \frac{\partial^2}{\partial t^2} \mathbf{E} = 0 \tag{79}$$

Since in the present case $\mathbf{E}(t)$ is transverse, i.e., perpendicular to the direction of propagation, the quantity $\nabla \times \nabla \times \mathbf{E}$ may be replaced by $-\nabla^2 \mathbf{E}$. Moreover, since $\mathbf{E}_0(t)$ is a slowly varying function in comparison to $e^{-i\omega t}$, we may replace $\partial/\partial t$ by $-i\omega$ and $\partial^2/\partial t^2$ by $-\omega^2$. Thus when expression (78) is substituted in Eq. (79) we find that the propagation constant is given by

$$k^2 = \omega^2 \mu_0 \left(\epsilon + i \frac{\sigma}{\omega} \right) \tag{80}$$

Since ω is assumed real, it is clear from Eq. (80) that k is generally

[1] See, for example, L. Brillouin, *Congr. intern. elec., Paris, 1932*, vol. 2, pp. 739–788, Gauthier-Villars, Paris, 1933.

complex. Accordingly, we write k in the form

$$k = \beta + i\alpha = \frac{\omega}{c}\eta + i\alpha \qquad (\alpha, \beta, \eta = \text{positive-definite}) \tag{81}$$

which displays as real quantities the phase factor β, the attenuation factor α, and the index of refraction η. To obtain explicit expressions for these factors in terms of the constitutive parameters, we substitute relations (81) into Eq. (80). Thus we find that

$$\beta = \omega\sqrt{\mu_0}\left[\frac{\epsilon}{2} + \sqrt{\left(\frac{\epsilon}{2}\right)^2 + \left(\frac{\sigma}{2\omega}\right)^2}\right]^{1/2} \tag{82}$$

$$\alpha = \omega\sqrt{\mu_0}\left[-\frac{\epsilon}{2} + \sqrt{\left(\frac{\epsilon}{2}\right)^2 + \left(\frac{\sigma}{2\omega}\right)^2}\right]^{1/2} \tag{83}$$

$$\eta = \frac{1}{\sqrt{\epsilon_0}}\left[\frac{\epsilon}{2} + \sqrt{\left(\frac{\epsilon}{2}\right)^2 + \left(\frac{\sigma}{2\omega}\right)^2}\right]^{1/2} \tag{84}$$

Applying expressions (82), (83), and (84) to a lossless (nonabsorptive) plasma whose constitutive parameters are $\epsilon = \epsilon_0(1 - \omega_p^2/\omega^2)$, $\mu = \mu_0$, $\sigma = 0$, we get

$$\beta = \frac{\omega}{c}\sqrt{1 - (\omega_p^2/\omega^2)} \qquad \alpha = 0$$

$$\eta = \sqrt{1 - (\omega_p^2/\omega^2)} \qquad \text{for } \omega > \omega_p \tag{85}$$

$$\beta = 0 \qquad \alpha = \frac{\omega}{c}\sqrt{\frac{\omega_p^2}{\omega^2} - 1} \qquad \eta = 0 \qquad \text{for } \omega < \omega_p \tag{86}$$

$$\beta = 0 \qquad \alpha = 0 \qquad \eta = 0 \qquad \text{for } \omega = \omega_p \tag{87}$$

These expressions show the marked difference in behavior between a wave whose operating frequency is greater than the plasma frequency and a wave whose operating frequency is less than the plasma frequency. When $\omega > \omega_p$, the wave travels without attenuation at a phase velocity greater than that of light in vacuum. On the other hand, when $\omega < \omega_p$ the wave is evanescent (nonabsorptively damped) and carries no power. At $\omega = \omega_p$ the wave is cut off; the magnetic field is zero and the electric field must satisfy $\nabla \times \mathbf{E}(t) = 0$. Hence, at cutoff a transverse electromagnetic wave cannot exist. However, a longitudinal electrical wave, sometimes called a "plasma wave" or "electrostatic wave," can exist. To examine the properties of such a wave, spatial dispersion must be taken into account.

There are three types of velocity that pertain to the transverse wave: the phase velocity v_{ph}, whose value can be found from a knowledge of η by using the relation $v_{ph} = c/\eta$; the group velocity v_g, which by definition is $\partial\omega/\partial\beta$; and the velocity of energy transport v_{en}, which is defined by the ratio $\bar{S}_z/\bar{w}$. Again restricting the discussion to a lossless plasma, we see from expressions (85) that the phase and group velocities are given by

$$v_{ph} \equiv \frac{\omega}{\beta} = \frac{c}{\sqrt{1 - (\omega_p{}^2/\omega^2)}} \qquad (\omega \geq \omega_p) \tag{88}$$

$$v_g \equiv \frac{\partial\omega}{\partial\beta} = c\sqrt{1 - \frac{\omega_p{}^2}{\omega^2}} \qquad (\omega \geq \omega_p) \tag{89}$$

Since an increase of wavelength (or, equivalently, a decrease of frequency) results in an increase in phase velocity, the dispersion is said to be "normal."

To find v_{en}, we note that the time-average value of the Poynting vector of the wave is z directed and has the value

$$\bar{S}_z = \tfrac{1}{2}\,\mathrm{Re}\;\mathbf{e}_z \cdot (\mathbf{E} \times \mathbf{H}^*) = \tfrac{1}{2}\,\mathrm{Re}\sqrt{\frac{\epsilon}{\mu_0}}\,\mathbf{E}_0 \cdot \mathbf{E}_0^* \tag{90}$$

Moreover, we note that the time-average energy density (77) in this case reduces to

$$\bar{w} = \tfrac{1}{4}\epsilon\mathbf{E}_0 \cdot \mathbf{E}_0^* + \frac{1}{4}\frac{\partial}{\partial\omega}(\omega\epsilon)\mathbf{E}_0 \cdot \mathbf{E}_0^* \tag{91}$$

Therefore, the velocity of energy transport assumes the form

$$v_{en} = \frac{\bar{S}_z}{\bar{w}} = \frac{(\frac{1}{2})\,\mathrm{Re}\sqrt{\epsilon/\mu_0}}{(\frac{1}{2})\epsilon + (\frac{1}{4})\omega\,\partial\epsilon/\partial\omega} \tag{92}$$

Substituting $\epsilon = \epsilon_0(1 - \omega_p^2/\omega^2)$ into this form, we find that the velocity of energy transport in a lossless plasma is given by

$$v_{en} = c\sqrt{1 - \frac{\omega_p^2}{\omega^2}} \qquad (\omega \geq \omega_p) \tag{93}$$

which is identical to expression (89) for the group velocity.

Let us consider now a plasma with small losses. In the limiting case where $|\epsilon| \gg \sigma/\omega$, the losses are incidental and expressions (82), (83), and (84) reduce to

$$\beta = \omega\sqrt{\mu_0\epsilon} \qquad \alpha = \frac{\sigma}{2}\sqrt{\frac{\mu_0}{\epsilon}} \qquad \eta = \sqrt{\epsilon/\epsilon_0} \qquad (\omega \geq \omega_p) \tag{94}$$

Using relations (44) and (45), i.e.,

$$\epsilon = \epsilon_0\left(1 - \frac{\omega_p^2}{\omega^2 + \omega_{eff}^2}\right) \qquad \sigma = \frac{\epsilon_0\omega_{eff}\omega_p^2}{\omega^2 + \omega_{eff}^2} \tag{95}$$

we see that expressions (94) yield

$$\beta = \frac{\omega}{c}\sqrt{1 - \frac{\omega_p^2}{\omega^2 + \omega_{eff}^2}} \tag{96}$$

$$\alpha = \frac{\frac{1}{2}\,\omega_{eff}\omega_p^2}{c(\omega^2 + \omega_{eff}^2)\sqrt{1 - \omega_p^2/(\omega^2 + \omega_{eff}^2)}} \tag{97}$$

$$\eta = \sqrt{1 - \frac{\omega_p^2}{\omega^2 + \omega_{eff}^2}} \tag{98}$$

and the corresponding phase and group velocities are given by

$$v_{ph} = \frac{c}{\sqrt{1 - \dfrac{\omega_p^2}{\omega^2 + \omega_{eff}^2}}} \tag{99}$$

$$v_g = \frac{c\sqrt{1 - \omega_p^2/(\omega^2 + \omega_{eff}^2)}}{1 - \omega_p^2\omega_{eff}^2/(\omega^2 + \omega_{eff}^2)^2} \tag{100}$$

Comparing expression (88) with expression (99), we see that the phase velocity is decreased by the presence of loss. On the other hand, comparing expression (89) with expression (100), we see that the group velocity is increased by the presence of loss. The interpretation of group velocity as the velocity of energy transport breaks down when the medium is dissipative.

6.5 Dielectric Tensor of Magnetically Biased Plasma

When a magnetostatic field $\mathbf{B}_0$ is applied to a plasma, the plasma becomes electrically anisotropic for electromagnetic waves. That is, the permeability of the plasma remains equal to the vacuum permeability μ_0, whereas the dielectric constant of the plasma is transformed into a tensor[1] (or dyadic) quantity $\boldsymbol{\varepsilon}$.

To derive the dielectric tensor of a magnetically biased plasma, which for simplicity is assumed for the present to be lossless, we use the macroscopic equation of motion (35). In the present instance this equation reduces to

$$-inm\omega\mathbf{v} = nq(\mathbf{E} + \mathbf{v} \times \mathbf{B}_0) \tag{101}$$

and yields the following expression for the macroscopic velocity of the plasma electrons:

$$\mathbf{v} = \frac{-\omega^2(q/m)\mathbf{E} - i\omega(q^2/m^2)\mathbf{E} \times \mathbf{B}_0 + (q^3/m^3)(\mathbf{E} \cdot \mathbf{B}_0)\mathbf{B}_0}{-i\omega\left[\left(\frac{q}{m}\mathbf{B}_0\right) \cdot \left(\frac{q}{m}\mathbf{B}_0\right) - \omega^2\right]} \tag{102}$$

Since the density of the electronic convection current $\mathbf{J}$ by definition is equal to $nq\mathbf{v}$, it follows from expression (102) that $\mathbf{J}$ is given by

$$\mathbf{J} = -i\omega \frac{\epsilon_0\omega_p^{\,2}}{(\boldsymbol{\omega}_g \cdot \boldsymbol{\omega}_g - \omega^2)} \mathbf{E} + \frac{\epsilon_0\omega_p^{\,2}}{(\boldsymbol{\omega}_g \cdot \boldsymbol{\omega}_g - \omega^2)} \mathbf{E} \times \boldsymbol{\omega}_g - \frac{1}{i\omega} \frac{\varepsilon_0\,\omega_p^{\,2}}{(\boldsymbol{\omega}_g \cdot \boldsymbol{\omega}_g - \omega^2)} (\mathbf{E} \cdot \boldsymbol{\omega}_g)\boldsymbol{\omega}_g \tag{103}$$

where ω_p is the plasma frequency ($\omega_p^{\,2} = nq^2/m\epsilon_0$) and where the

[1] See, for example, C. H. Papas, A Note Concerning a Gyroelectric Medium, *Caltech Tech. Rept.* 4, prepared for the Office of Naval Research, May, 1954.

amplitude[1] ω_g of the vector

$$\boldsymbol{\omega}_g \equiv \frac{q}{m}\mathbf{B}_0 \tag{104}$$

represents the gyrofrequency of the electrons. From a knowledge of **J** we can find the dielectric constant of the plasma by noting that the total current density is the sum of the convection current density **J** and the vacuum displacement current density $-i\omega\epsilon_0\mathbf{E}$, and then by regarding this total current density as a displacement current in a dielectric medium whose dielectric constant $\boldsymbol{\varepsilon}$ is fixed by the relation

$$\mathbf{J} - i\omega\epsilon_0\mathbf{E} = -i\omega\boldsymbol{\varepsilon}\cdot\mathbf{E} \tag{105}$$

According to expression (103), it appears that **J** is generally not parallel to **E**; the quantity $\boldsymbol{\varepsilon}$ must be a tensor or dyadic to take this into account. Since, by definition, the displacement vector **D** is calculated from

$$\mathbf{D} = \boldsymbol{\varepsilon}\cdot\mathbf{E} \tag{106}$$

the tensor character of $\boldsymbol{\varepsilon}$ also means that **D** is not generally parallel to **E**.

Although a tensor is independent of coordinates, its components are not. If we are given the components of a tensor with respect to one coordinate system, we can find its components with respect to any other coordinate system by applying the transformation law connecting the coordinates of one system with those of the other. Therefore we are free to choose any coordinate system without risking loss of generality. In the present instance, for simplicity, we choose a cartesian system of coordinates (x,y,z) whose z axis is parallel to $\mathbf{B}_0$, that is, $\mathbf{B}_0 = \mathbf{e}_z B_0$; $\mathbf{e}_z$ is the z-directed unit vector. When $B_0 > 0$, the vector $\mathbf{B}_0$ is parallel to the z axis; and when $B_0 < 0$, the vector $\mathbf{B}_0$ is antiparallel to the z axis. The components of $\boldsymbol{\varepsilon}$ in this cartesian system are denoted by ϵ_{ik}, with $i, k = x, y, z$.

Substituting expression (103) into Eq. (105) leads to the following expressions for the components ϵ_{ik} of $\boldsymbol{\varepsilon}$ in the cartesian system whose

[1] This means that $\omega_g = (q/m)B_0$. For electrons q is negative and hence $\omega_g = (-|q|/m)B_0$.

z axis is parallel to $\mathbf{B}_0$:

$$\epsilon_{xx} = \epsilon_0 \left(1 - \frac{\omega_p{}^2}{\omega^2 - \omega_g{}^2}\right) = \epsilon_{yy} \tag{107}$$

$$\epsilon_{xy} = -i\epsilon_0 \frac{\omega_p{}^2\omega_g}{\omega(\omega^2 - \omega_g{}^2)} = -\epsilon_{yx} \tag{108}$$

$$\epsilon_{zz} = \epsilon_0 \left(1 - \frac{\omega_p{}^2}{\omega^2}\right) \tag{109}$$

The remaining components ϵ_{xz}, ϵ_{zx}, ϵ_{yz}, ϵ_{zy} are identically zero. We note that when the magnetostatic field B_0 vanishes, ω_g vanishes and the diagonal terms become equal to each other, i.e.,

$$\epsilon_{xx} = \epsilon_{yy} = \epsilon_{zz} = \epsilon_0(1 - \omega_p{}^2/\omega^2) \tag{110}$$

and the off-diagonal terms disappear. That is, when $B_0 = 0$, the plasma becomes isotropic as it should. Also we note that when $\mathbf{B}_0$ is replaced by $-\mathbf{B}_0$, the gyrofrequency ω_g changes sign and, consequently, the components satisfy the generalized symmetry relation

$$\epsilon_{ik}(\mathbf{B}_0) = \epsilon_{ki}(-\mathbf{B}_0) \tag{111}$$

as must the components of the dielectric constant of any medium whose anisotropy is due to an externally applied magnetostatic field.[1] In addition we see that the components constitute a hermitian matrix, i.e.,

$$\epsilon_{ik}(\mathbf{B}_0) = \epsilon_{ki}^*(\mathbf{B}_0) \tag{112}$$

The hermitian nature of the dielectric tensor results from the assumption that the plasma is lossless.

Expressions (107), (108), and (109) for the components of the dielectric tensor may be easily generalized to take into account collision losses. For the case where the collision losses are appreciable, we must add to the right side of Eq. (101) a collision term. Thus for the equa-

[1] See, for example, A. Sommerfeld, "Lectures on Theoretical Physics," vol. 5, "Thermodynamics and Statistical Mechanics," p. 163, Academic Press Inc., New York, 1956.

tion of motion of the electrons we get

$$-inm\omega\mathbf{v} = nq(\mathbf{E} + \mathbf{v} \times \mathbf{B}_0) - nm\mathbf{v}\omega_{eff} \tag{113}$$

where ω_{eff} is the collision frequency. Rewriting this equation in the form

$$-inm(\omega + i\omega_{eff})\mathbf{v} = nq(\mathbf{E} + \mathbf{v} \times \mathbf{B}_0) \tag{114}$$

and comparing with Eq. (101), we see that the resulting expression for $\mathbf{J}$ is the same as expression (103), with ω replaced by $\omega + i\omega_{eff}$. It then follows from Eq. (105) that the cartesian components of the dielectric tensor of a lossy dielectric are given by

$$\epsilon'_{xx} = \epsilon_0\left(1 - \frac{\omega_p{}^2(\omega + i\omega_{eff})}{\omega[(\omega + i\omega_{eff})^2 - \omega_g{}^2]}\right) = \epsilon'_{yy} \tag{115}$$

$$\epsilon'_{xy} = -i\epsilon_0 \frac{\omega_p{}^2\omega_g}{\omega(\omega + i\omega_{eff} + \omega_g)(\omega + i\omega_{eff} - \omega_g)} = -\epsilon'_{yx} \tag{116}$$

$$\epsilon'_{zz} = \epsilon_0\left[1 - \frac{\omega_p{}^2}{(\omega\omega + i\omega_{eff})}\right] \tag{117}$$

where the prime is used to distinguish the lossy components from the loss-free ones. As in the lossless case, we again have

$$\epsilon'_{ik}(\mathbf{B}_0) = \epsilon'_{ki}(-\mathbf{B}_0) \tag{118}$$

but, unlike the lossless case, the components ϵ'_{ik} do not constitute a hermitian matrix. We can, however, decompose ϵ'_{ik} uniquely as follows,

$$\epsilon'_{ik} = \epsilon_{ik} + \frac{i}{\omega}\sigma_{ik} \tag{119}$$

so that ϵ_{ik} and σ_{ik} are hermitian.

When the frequency of the electromagnetic waves that are passing through a magnetically biased plasma is very low, the motion of the plasma ions must be included in the analysis. We can find the dielectric constant in this low-frequency case by calculating the convection current as the sum of the ionic current and the previously determined electronic current, and by finding $\boldsymbol{\varepsilon}$ from a knowledge of $\mathbf{J}$ through the use of relation (105).

To proceed with the calculation, we note that the equation of motion for the ions is formally the same as the equation of motion for the electrons. Accordingly, since we have (in the loss-free case)

$$-inm\omega\mathbf{v} = nq(\mathbf{E} + \mathbf{v} \times \mathbf{B}_0) \tag{120}$$

as the equation of motion for the electrons, then for the ions the equation of motion must be

$$-in_i m_i \omega \mathbf{v}_i = n_i q_i(\mathbf{E} + \mathbf{v}_i \times \mathbf{B}_0) \tag{121}$$

Here m_i denotes the ionic mass, q_i the ionic charge, n_i the ionic population density, and $\mathbf{v}_i$ the macroscopic velocity of the ions. We know from previous calculation that the electronic convection current $nq\mathbf{v}$ is given by expression (103). Hence, it follows from the similarity of Eqs. (120) and (121) that the ionic convection current $n_i q_i \mathbf{v}_i$ is given by the same expression (103) but with ω_p replaced by the ionic plasma frequency ω_{pi} and ω_g replaced by the ionic gyrofrequency, where

$$\omega_{pi}{}^2 = \frac{n_i q_i{}^2}{m_i \epsilon_0} \qquad \omega_{gi} = \frac{q_i}{m_i}\mathbf{B}_0 \tag{122}$$

Superposing $nq\mathbf{v}$ and $n_i q_i \mathbf{v}_i$, we get $\mathbf{J}$, that is,

$$\mathbf{J} = nq\mathbf{v} + n_i q_i \mathbf{v}_i \tag{123}$$

and then substituting this $\mathbf{J}$ into relation (105), we find that the nonzero components of $\boldsymbol{\varepsilon}$ for a loss-free magnetically biased plasma are given by[1]

$$\epsilon_{xx} = \epsilon_0\left(1 - \frac{\omega_p{}^2}{\omega^2 - \omega_g{}^2} - \frac{\omega_{pi}{}^2}{\omega^2 - \omega_{gi}{}^2}\right) = \epsilon_{yy} \tag{124}$$

$$\epsilon_{xy} = -i\epsilon_0\left[\frac{\omega_p{}^2\omega_g}{\omega(\omega^2 - \omega_g{}^2)} + \frac{\omega_{pi}{}^2\omega_{gi}}{\omega(\omega^2 - \omega_{gi}{}^2)}\right] = -\epsilon_{yx} \tag{125}$$

$$\epsilon_{zz} = \epsilon_0\left(1 - \frac{\omega_p{}^2}{\omega^2} - \frac{\omega_{pi}{}^2}{\omega^2}\right) \tag{126}$$

[1] See, for example, E. Astrom, On Waves in an Ionized Gas, *Arkiv Fysik*, **2:** 443 (1950).

These components are in accord with the generalized symmetry relation (111) and with the hermiticity condition (112).

The hermitian property of the dielectric tensor is a consequence of the assumption that the plasma is loss-free. To show that the hermiticity of the tensor is preserved under a rotation of the coordinate system, we introduce another cartesian system x', y', z', which is obtained from the original cartesian system x, y, z by a pure rotation. Let $\mathbf{a}_i$, with $i = x', y', z'$, denote the unit vectors along the axes of the primed system, and as before let $\mathbf{e}_i$, with $i = x, y, z$, denote the unit vectors along the axes of the unprimed system. In the unprimed system the dielectric tensor is given by

$$\boldsymbol{\varepsilon} = \Sigma \mathbf{e}_i \mathbf{e}_k \epsilon_{ik} \qquad (i, k = x, y, z) \tag{127}$$

and in the primed system it must have the form

$$\boldsymbol{\varepsilon} = \Sigma \mathbf{a}_i \mathbf{a}_k \epsilon_{ik}{}^{(a)} \qquad (i, k = x', y', z') \tag{128}$$

where $\epsilon_{ik}{}^{(a)}$ denote the components of $\boldsymbol{\varepsilon}$ with respect to the primed system. Since $\mathbf{a}_i \cdot \mathbf{a}_k = \delta_{ik}$, it follows from expression (128) that

$$\epsilon_{mn}{}^{(a)} = \mathbf{a}_m \cdot \boldsymbol{\varepsilon} \cdot \mathbf{a}_n \tag{129}$$

Substituting expression (127) in expression (129), we obtain the relation

$$\epsilon_{mn}{}^{(a)} = \Sigma(\mathbf{a}_m \cdot \mathbf{e}_i)(\mathbf{e}_k \cdot \mathbf{a}_n)\epsilon_{ik} \tag{130}$$

which, by means of the shorthand

$$\gamma_{ik} \equiv \mathbf{a}_i \cdot \mathbf{e}_k \tag{131}$$

can be written as

$$\epsilon_{mn}{}^{(a)} = \Sigma \gamma_{mi} \gamma_{nk} \epsilon_{ik} \tag{132}$$

Similarly we obtain

$$\epsilon_{nm}{}^{(a)} = \Sigma \gamma_{ni} \gamma_{mk} \epsilon_{ik} \tag{133}$$

Since ϵ_{ik} is hermitian, it follows from Eqs. (132) and (133) that $\epsilon_{nm}{}^{(a)}$ is also hermitian, i.e.,

$$\epsilon_{mn}{}^{(a)} = \epsilon_{nm}{}^{(a)*} \tag{134}$$

Thus we see that hermiticity is preserved under a rotation of the axes.

Although we have found it convenient to express the constitutive relation of a magnetically biased plasma by means of a single tensor, it is simpler in certain considerations to deal instead with the elementary vector operations that carry **E** into **D**. To determine these operations, we assume for simplicity that the plasma is loss-free. Consequently, its dielectric tensor has the form

$$\boldsymbol{\varepsilon} = \begin{pmatrix} a & -ig & 0 \\ ig & a & 0 \\ 0 & 0 & b \end{pmatrix} \tag{135}$$

where a, b, and g are real quantities. Splitting this matrix as follows,

$$\epsilon = \begin{pmatrix} a & 0 & 0 \\ 0 & a & 0 \\ 0 & 0 & a \end{pmatrix} + \begin{pmatrix} 0 & 0 & 0 \\ 0 & 0 & 0 \\ 0 & 0 & b-a \end{pmatrix} + \begin{pmatrix} 0 & -ig & 0 \\ ig & 0 & 0 \\ 0 & 0 & 0 \end{pmatrix} \tag{136}$$

and then substituting it into the constitutive relation $\mathbf{D} = \boldsymbol{\varepsilon} \cdot \mathbf{E}$, we obtain

$$\mathbf{D} = a\mathbf{E} + (b-a)\mathbf{e}_z(\mathbf{e}_z \cdot \mathbf{E}) + ig\mathbf{e}_z \times \mathbf{E} \tag{137}$$

as an alternative statement of the constitutive relation. In the case where the motion of the ions can be neglected, i.e., in the case where a, b, and ig are given respectively by expressions (107), (109), and (108), we have

$$\mathbf{D} = \epsilon_0\left(1 - \frac{\omega_p{}^2}{\omega^2 - \omega_g{}^2}\right)\mathbf{E} + \epsilon_0 \frac{\omega_p{}^2}{\omega^2(\omega^2 - \omega_g{}^2)}\,\boldsymbol{\omega}_g(\boldsymbol{\omega}_g \cdot \mathbf{E}) + i\epsilon_0 \frac{\omega_p{}^2}{\omega(\omega^2 - \omega_g{}^2)}\,\boldsymbol{\omega}_g \times \mathbf{E} \tag{138}$$

When the biasing field is weak or when the frequency is high, the ratio ω_g/ω is small compared to unity and relation (138) to first order in

ω_g/ω becomes

$$\mathbf{D} = \epsilon\mathbf{E} + i\epsilon_0 \frac{\omega_p^2}{\omega^3} \boldsymbol{\omega}_g \times \mathbf{E} \tag{139}$$

where $\epsilon = \epsilon_0(1 - \omega_p^2/\omega^2)$ is the dielectric constant of an isotropic plasma.

Returning to the dielectric tensor (135), we ask whether there is a special coordinate system with respect to which the dielectric tensor is diagonal. The answer to this is that since the tensor is hermitian its matrix can be diagonalized by a unitary transformation which amounts to a complex rotation in Hilbert space.[1] More simply, however, we observe that when the dielectric tensor (135) is substituted into the constitutive relation $\mathbf{D} = \boldsymbol{\varepsilon} \cdot \mathbf{E}$, we obtain

$$D_x = aE_x - igE_y \tag{140}$$

$$D_y = igE_x + aE_y \tag{141}$$

$$D_z = bE_z \tag{142}$$

With the aid of the following combinations of expressions (140) and (141),

$$D_x + iD_y = (a - g)(E_x + iE_y) \tag{143}$$

$$D_x - iD_y = (a + g)(E_x - iE_y) \tag{144}$$

we get the matrix equation

$$\begin{pmatrix} D_x + iD_y \\ D_x - iD_y \\ D_z \end{pmatrix} = \begin{pmatrix} a - g & 0 & 0 \\ 0 & a + g & 0 \\ 0 & 0 & b \end{pmatrix} \begin{pmatrix} E_x + iE_y \\ E_x - iE_y \\ E_z \end{pmatrix} \tag{145}$$

which displays the dielectric constant as a diagonal matrix. Since $\mathbf{D} \cdot (\mathbf{e}_x \pm i\mathbf{e}_y) = (\mathbf{e}_x D_x + \mathbf{e}_y D_y + \mathbf{e}_z D_z) \cdot (\mathbf{e}_x \pm i\mathbf{e}_y) = D_x \pm iD_y$, the component $D_x + iD_y$ is the projection of $\mathbf{D}$ on the vector $\mathbf{e}_x + i\mathbf{e}_y$ and

[1] See, for example, Hermann Weyl, "The Theory of Groups and Quantum Mechanics," chap. 1, Dover Publications, Inc., 1931. Translated from the German by H. P. Robertson.

$D_x - iD_y$ is the projection of **D** on the vector $\mathbf{e}_x - i\mathbf{e}_y$. Thus we see that the elements of the matrices in Eq. (145) are referred to a coordinate system $x' = 1/\sqrt{2}\,(x + iy)$, $y' = 1/\sqrt{2}\,(x - iy)$, $z' = z$, whose unit vectors are $\mathbf{e}_{x'} = 1/\sqrt{2}\,(\mathbf{e}_x + i\mathbf{e}_y)$, $\mathbf{e}_{y'} = 1/\sqrt{2}\,(\mathbf{e}_x - i\mathbf{e})_y$, $\mathbf{e}_{z'} = \mathbf{e}_z$. The vectors $\mathbf{e}_{x'}$, $\mathbf{e}_{y'}$, $\mathbf{e}_{z'}$, which are unit orthogonal vectors in the hermitian sense, that is, $\mathbf{e}_i \cdot \mathbf{e}_k^* = \delta_{ik}$ where $i, k = x', y', z'$, constitute the principal axes of the dielectric tensor.[1]

6.6 Plane Wave in Magnetically Biased Plasma

In this section we shall study the propagation and polarization properties of a plane monochromatic wave in a magnetically biased homogeneous plasma which for simplicity is assumed to be lossless. We regard the plasma as a continuous medium whose conductivity is zero, whose permeability is equal to the vacuum permeability μ_0, and whose dielectric constant is the tensor $\boldsymbol{\varepsilon}$ given by Eqs. (107), (108), and (109) of the previous section.

By definition, the electric vector of a plane monochromatic wave has the form

$$\mathbf{E}(\mathbf{r}) = \mathbf{E}_0 e^{i\mathbf{k}\cdot\mathbf{r}} \tag{146}$$

where $\mathbf{E}_0$ is a constant vector, **k** is the vector wave number, and **r** is the position vector. We may write **k** as

$$\mathbf{k} = \mathbf{n}\frac{\omega}{v} \tag{147}$$

where **n** is the unit vector in the direction of propagation and v is the phase velocity of the wave. The problem is to determine the vector **k**, which describes the propagation of the wave, and the vector $\mathbf{E}_0$, which describes the polarization of the wave.

[1] For an exhaustive discussion, see G. Lange-Hesse, Vergleich der Doppelbrechung in Kristall und in der Ionosphäre, *Archiv der Elektrischen Übertragung*, **6:** 149–158 (1952).

The vector **E** must satisfy the Helmholtz equation

$$\nabla \times \nabla \times \mathbf{E} = \omega^2 \mu_0 \boldsymbol{\varepsilon} \cdot \mathbf{E} \tag{148}$$

as can be seen from the Maxwell equations

$$\nabla \times \mathbf{E} = i\omega\mu_0 \mathbf{H} \qquad \nabla \times \mathbf{H} = -i\omega\boldsymbol{\varepsilon} \cdot \mathbf{E} \tag{149}$$

by taking the curl of the first and then using the second to eliminate **H**. Substituting expression (146) into Eq. (148), and using relation (147), we obtain

$$\mathbf{E}_0 - \mathbf{n}(\mathbf{n} \cdot \mathbf{E}_0) = \frac{1}{\epsilon_0} \frac{v^2}{c^2} \boldsymbol{\varepsilon} \cdot \mathbf{E}_0 \tag{150}$$

where $c = 1/\sqrt{\mu_0 \epsilon_0}$ is the vacuum velocity of light. Without loss of generality we choose a cartesian system of coordinates so oriented that the z axis is parallel to $\mathbf{B}_0$ and the yz plane contains $\mathbf{n}$. As shown in Fig. 6.1, the angle between $\mathbf{n}$ and $\mathbf{B}_0$ is denoted by θ. Accordingly, the x, y, z components of the vector equation (150) are given by

$$\begin{aligned}
&E_{0x}\left(1 - \frac{v^2}{c^2}\frac{\epsilon_{xx}}{\epsilon_0}\right) - E_{0y}\left(\frac{v^2}{c^2}\frac{\epsilon_{xy}}{\epsilon_0}\right) + 0 = 0 \\
&E_{0x}\left(-\frac{v^2}{c^2}\frac{\epsilon_{yx}}{\epsilon_0}\right) + E_{0y}\left(\cos^2\theta - \frac{v^2}{c^2}\frac{\epsilon_{yy}}{\epsilon_0}\right) + E_{0z}(-\cos\theta\sin\theta) = 0 \\
&0 + E_{0y}(-\cos\theta\sin\theta) + E_{0z}\left(\sin^2\theta - \frac{v^2}{c^2}\frac{\epsilon_{zz}}{\epsilon_0}\right) = 0
\end{aligned} \tag{151}$$

where E_{0x}, E_{0y}, E_{0z} are the cartesian components of $\mathbf{E}_0$. Since these three simultaneous equations are homogeneous, they yield a nontrivial solution only when

$$\begin{vmatrix}
\left(1 - \frac{v^2}{c^2}\frac{\epsilon_{xx}}{\epsilon_0}\right) & -\frac{v^2}{c^2}\frac{\epsilon_{xy}}{\epsilon_0} & 0 \\
-\frac{v^2}{c^2}\frac{\epsilon_{yx}}{\epsilon_0} & \left(\cos^2\theta - \frac{v^2}{c^2}\frac{\epsilon_{yy}}{\epsilon_0}\right) & -\sin\theta\cos\theta \\
0 & -\sin\theta\cos\theta & \sin^2\theta - \frac{v^2}{c^2}\frac{\epsilon_{zz}}{\epsilon_0}
\end{vmatrix} = 0 \tag{152}$$

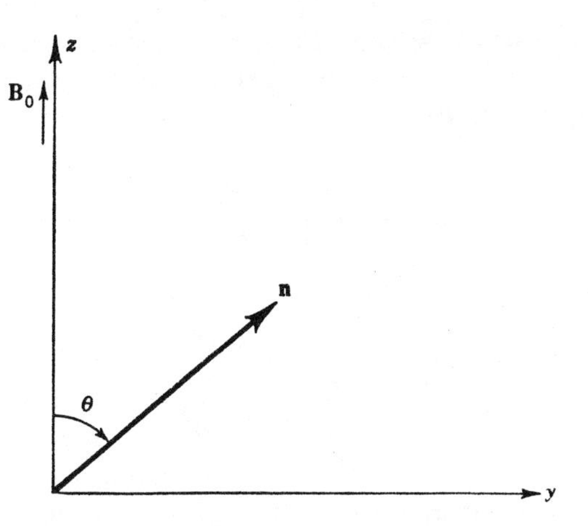

Fig. 6.1 Arbitrary direction **n** *of wave propagation in plasma with applied magnetostatic field* $\mathbf{B}_0$.

With the aid of the quantities ϵ_1, ϵ_2, ϵ_3, which are defined by

$$\epsilon_1 = \frac{\epsilon_{xx}}{\epsilon_0} - i\frac{\epsilon_{xy}}{\epsilon_0} \qquad \epsilon_2 = \frac{\epsilon_{xx}}{\epsilon_0} + i\frac{\epsilon_{xy}}{\epsilon_0} \qquad \epsilon_3 = \frac{\epsilon_{zz}}{\epsilon_0} \tag{153}$$

we find that Eq. (152) can be written

$$-\tan^2\theta = \frac{\left(\dfrac{v^2}{c^2} - \dfrac{1}{\epsilon_1}\right)\left(\dfrac{v^2}{c^2} - \dfrac{1}{\epsilon_2}\right)}{\left(\dfrac{v^2}{c^2} - \dfrac{1}{\epsilon_3}\right)\left[\dfrac{v^2}{c^2} - \dfrac{1}{2}\left(\dfrac{1}{\epsilon_1} + \dfrac{1}{\epsilon_2}\right)\right]} \tag{154}$$

This equation determines two values of v^2/c^2 for each value of θ.

In the case where the propagation is parallel to $\mathbf{B}_0$, we have $\theta = 0$; accordingly, Eq. (154) yields the two solutions

$$\frac{v^2}{c^2} = \frac{1}{\epsilon_1} = \frac{1}{\dfrac{\epsilon_{xx}}{\epsilon_0} - i\dfrac{\epsilon_{xy}}{\epsilon_0}} = \frac{1}{1 - \dfrac{X}{1+Y}} \tag{155}$$

and

$$\frac{v^2}{c^2} = \frac{1}{\epsilon_2} = \frac{1}{\dfrac{\epsilon_{xx}}{\epsilon_0} + i\dfrac{\epsilon_{xy}}{\epsilon_0}} = \frac{1}{1 - \dfrac{X}{1-Y}} \tag{156}$$

where $X = (\omega_p/\omega)^2$ and $Y = -\omega_g/\omega$.[1] From these expressions it follows that the propagation constants of the two waves that travel parallel to $\mathbf{B}_0$ are given by

$$k_0' = \frac{\omega}{c}\sqrt{1 - \frac{X}{1+Y}} = \frac{\omega}{c}\sqrt{1 - \frac{\omega_p{}^2}{\omega(\omega - \omega_g)}} \tag{157}$$

and

$$k_0'' = \frac{\omega}{c}\sqrt{1 - \frac{X}{1-Y}} = \frac{\omega}{c}\sqrt{1 - \frac{\omega_p{}^2}{\omega(\omega + \omega_g)}} \tag{158}$$

Moreover, when the propagation is along the y axis, i.e., perpendicular to $\mathbf{B}_0$, θ is equal to $\pi/2$ and in this case the two solutions of Eq. (154) are

$$\frac{v^2}{c^2} = \frac{1}{\epsilon_3} = \frac{\epsilon_0}{\epsilon_{zz}} = \frac{1}{1 - X} \tag{159}$$

and

$$\frac{v^2}{c^2} = \frac{1}{2}\left(\frac{1}{\epsilon_1} + \frac{1}{\epsilon_2}\right) = \frac{1}{1 - \dfrac{X}{1 - Y^2/(1 - X)}} \tag{160}$$

For the propagation constants of the corresponding two waves, we have

$$k_{\pi/2}' = \frac{\omega}{c}\sqrt{1 - X} = \frac{\omega}{c}\sqrt{1 - \frac{\omega_p{}^2}{\omega^2}} \tag{161}$$

and

$$k_{\pi/2}'' = \frac{\omega}{c}\sqrt{1 - \frac{X}{1 - Y^2/(1 - X)}} = \frac{\omega}{c}\sqrt{1 - \frac{\omega_p{}^2/\omega^2}{1 - \omega_g{}^2/(\omega^2 - \omega_p{}^2)}} \tag{162}$$

[1] Since q in the case of electrons is a negative quantity, then ω_g, which is given by $(q/m)B_0$, is also a negative quantity. We wish Y to be a positive quantity and therefore we include a minus sign in the definition.

In general, when θ is arbitrary we have the two solutions

$$\frac{v^2}{c^2} = \left[1 - \frac{X}{1 - \frac{1}{2}\frac{Y_T^2}{1-X} \pm \sqrt{\frac{1}{4}\frac{Y_T^4}{(1-X)^2} + Y_L^2}}\right]^{-1} \tag{163}$$

and hence

$$k'_\theta = \frac{\omega}{c}\left[1 - \frac{X}{1 - \frac{1}{2}\frac{Y_T^2}{1-X} + \sqrt{\frac{1}{4}\frac{Y_T^4}{(1-X)^2} + Y_L^2}}\right]^{1/2} \tag{164}$$

$$k''_\theta = \frac{\omega}{c}\left[1 - \frac{X}{1 - \frac{1}{2}\frac{Y_T^2}{1-X} - \sqrt{\frac{1}{4}\frac{Y_T^4}{(1-X)^2} + Y_L^2}}\right]^{1/2} \tag{165}$$

where $Y_T = Y \sin\theta$ and $Y_L = Y \cos\theta$.

Thus we see that there are two waves traveling in any arbitrary direction θ, and that one of them has a propagation constant k'_θ given by expression (164) while the other has a propagation constant k''_θ given by expression (165). Since as a function of X the propagation constant k'_θ resembles the propagation constant of a wave in an isotropic plasma more closely than k''_θ does, the wave whose propagation constant is k'_θ is sometimes referred to as the ordinary wave and the wave whose propagation constant is k''_θ as the extraordinary wave. Indeed $k'_{\pi/2}$, the value k'_θ has when $\theta = \pi/2$, is identically equal to the propagation constant of a wave in an isotropic plasma.

Each of the field vectors of a wave is proportional to $\exp(i\mathbf{k}\cdot\mathbf{r})$. Therefore the Maxwell equations $\nabla \times \mathbf{E} = i\omega\mu\mathbf{H}$, $\nabla \times \mathbf{H} = -i\omega\mathbf{D}$ reduce to the relations

$$i\mathbf{k} \times \mathbf{E} = i\omega\mu_0\mathbf{H} \qquad i\mathbf{k} \times \mathbf{H} = -i\omega\mathbf{D} \tag{166}$$

which clearly indicate that the vectors $\mathbf{k}$, $\mathbf{E}$, $\mathbf{D}$ lie in a plane perpendicular to $\mathbf{H}$ (Fig. 6.2). Since $\mathbf{H}$ is necessarily perpendicular to $\mathbf{k}$, the wave cannot be an H wave (also known as a TE wave). In general the wave must be an E wave (also known as a TM wave), but in certain special directions the wave is a TEM wave. The Poynting vector

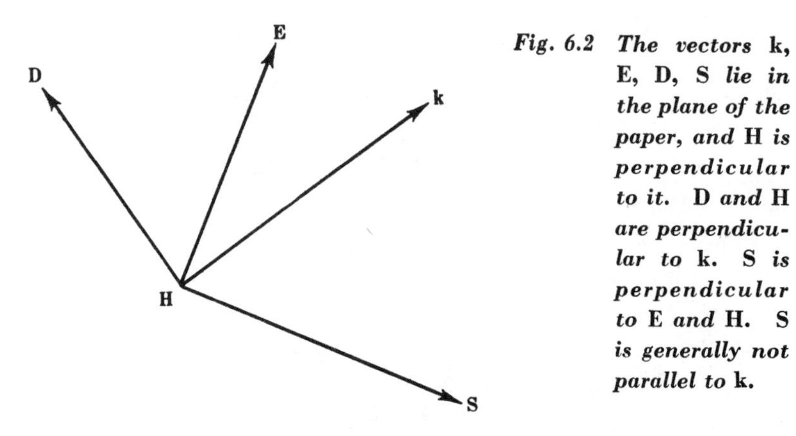

Fig. 6.2 ***The vectors* k, E, D, S *lie in the plane of the paper, and* H *is perpendicular to it.* D *and* H *are perpendicular to* k. S *is perpendicular to* E *and* H. S *is generally not parallel to* k.**

$\mathbf{S}(= \frac{1}{2}\mathbf{E} \times \mathbf{H}^*)$ of the wave is not parallel to $\mathbf{k}$ except in those directions of travel where the wave is TEM.[1]

Let us again consider the special case where the propagation is parallel to $\mathbf{B}_0$. In this case, $\theta = 0$ and Eqs. (151) reduce to

$$
\begin{aligned}
&E_{0x}\left(1 - \frac{v^2}{c^2}\frac{\epsilon_{xx}}{\epsilon_0}\right) - E_{0y}\left(\frac{v^2}{c^2}\frac{\epsilon_{xy}}{\epsilon_0}\right) = 0 \\
&E_{0x}\left(-\frac{v^2}{c^2}\frac{\epsilon_{yx}}{\epsilon_0}\right) + E_{0y}\left(1 - \frac{v^2}{c^2}\frac{\epsilon_{yy}}{\epsilon_0}\right) = 0 \\
&E_{0z}\left(-\frac{v^2}{c^2}\frac{\epsilon_{zz}}{\epsilon_0}\right) = 0
\end{aligned}
\tag{167}
$$

with v^2/c^2 given by Eq. (155) and by Eq. (156). From the third of these equations, we see that E_{0z} is zero. Consequently, the two waves that travel parallel to $\mathbf{B}_0$ are TEM waves. When v^2/c^2 is given by Eq. (155), the first or second of Eqs. (167) yields

$$\frac{E_{0x}}{E_{0y}} = i \tag{168}$$

and when v^2/c^2 is given by Eq. (156), we find that

$$\frac{E_{0x}}{E_{0y}} = -i \tag{169}$$

[1] However, it has been shown by S. M. Rytov, *J. Exptl. Theoret. Phys., U.S.S.R.*, **17:** 930 (1947), that the time-average Poynting vector is parallel to the group velocity.

Therefore, the electric vectors of the two waves traveling parallel to $\mathbf{B}_0$ can be written as

$$\mathbf{E}' = (\mathbf{e}_x - i\mathbf{e}_y)Ae^{ik_0'z} \tag{170}$$

and

$$\mathbf{E}'' = (\mathbf{e}_x + i\mathbf{e}_y)Ce^{ik_0''z} \tag{171}$$

where A and C are arbitrary amplitudes. Clearly $\mathbf{E}'$ is a left-handed circularly polarized wave, whereas $\mathbf{E}''$ is a right-handed circularly polarized wave.[1] The sum of these two waves yields the composite wave

$$\mathbf{E} = \mathbf{E}' + \mathbf{E}'' = \mathbf{e}_x(Ae^{ik_0'z} + Ce^{ik_0''z}) + \mathbf{e}_y(-iAe^{ik_0'z} + iCe^{ik_0''z}) \tag{172}$$

To study the polarization of this composite wave, we consider the ratio E_x/E_y. From (172) we obtain

$$\frac{E_x}{E_y} = i\,\frac{1 + (C/A)\exp\left[i(k_0'' - k_0')z\right]}{1 - (C/A)\exp\left[i(k_0'' - k_0')z\right]} \tag{173}$$

If the waves $\mathbf{E}'$ and $\mathbf{E}''$ are chosen to have equal amplitudes, then the constants A and C become equal. As a consequence of this choice, Eq. (173) reduces to

$$\frac{E_x}{E_y} = \cot\left(\frac{k_0' - k_0''}{2}\,z\right) \tag{174}$$

Since this relation is real, the composite wave at any position z is linearly polarized; however, the orientation angle of its plane of polarization (the plane containing $\mathbf{E}$ and $\mathbf{k}$) depends on z and rotates as z

[1] A geometric interpretation may be obtained by considering the real vectors $\text{Re}\,\mathbf{E}'e^{-i\omega t}$ and $\text{Re}\,\mathbf{E}''e^{-i\omega t}$. Setting $A = C = 1$, we obtain from Eqs. (170) and (171) the expressions

$$\text{Re}\,\mathbf{E}'e^{-i\omega t} = \mathbf{e}_x\cos(k_0'z - \omega t) + \mathbf{e}_y\sin(k_0'z - \omega t)$$

$$\text{Re}\,\mathbf{E}''e^{-i\omega t} = \mathbf{e}_x\cos(k_0''z - \omega t) - \mathbf{e}_y\sin(k_0''z - \omega t)$$

Clearly, at any fixed time the locus of the tip of the vector $\text{Re}\,\mathbf{E}'e^{-i\omega t}$ is a right-handed helix. As time increases this helix rotates counter-clockwise. On the other hand, the locus of the tip of the vector $\text{Re}\,\mathbf{E}''e^{-i\omega t}$ is a left-handed helix, which rotates clockwise.

increases or decreases. In other words, the composite wave undergoes Faraday rotation. The angle τ through which the resultant vector **E** rotates as the wave travels a unit distance is given by

$$\tau = \frac{k_0' - k_0''}{2} \tag{175}$$

The rotation is clockwise because $k_0' > k_0''$ always. With the aid of expressions (157) and (158), we see that τ can be written in the form[1]

$$\tau = \frac{1}{2}\frac{\omega}{c}\left[\sqrt{1 - \frac{\omega_p{}^2}{\omega(\omega - \omega_g)}} - \sqrt{1 - \frac{\omega_p{}^2}{\omega(\omega + \omega_g)}}\right] \tag{176}$$

which displays the dependence of the Faraday rotation τ on frequency.

We note that if a wave travels parallel to $\mathbf{B}_0$ it undergoes a clockwise Faraday rotation. On the other hand, if a wave travels antiparallel to $\mathbf{B}_0$ it undergoes a Faraday rotation of the opposite sense. That is, on reversing the direction of propagation, a clockwise wave becomes counterclockwise, and vice versa. This means that if the plane of polarization of a wave traveling parallel to $\mathbf{B}_0$ is rotated through a certain angle, then upon reflection it will be rotated still further, the rotation for the round trip being double the rotation for a single crossing.

For weak biasing fields the Faraday rotation depends linearly on B_0. To deduce this fact from expression (176), which in terms of the parameters $X = (\omega_p/\omega)^2$ and $Y = -\omega_g/\omega$ can be written as

$$\tau = \frac{1}{2}\frac{\omega}{c}\left(\sqrt{1 - \frac{X}{1 + Y}} - \sqrt{1 - \frac{X}{1 - Y}}\right) \tag{177}$$

we expand the square roots and retain only the first two terms in accord with the assumption that $X \ll 1$ and $Y \ll 1$. Thus we obtain the relation

$$\tau = \frac{1}{2}\frac{\omega}{c}XY = -\frac{1}{2c}\left(\frac{\omega_p}{\omega}\right)^2\omega_g \tag{178}$$

which shows that the Faraday rotation τ for weak biasing fields ($Y \ll 1$) and high frequencies ($X \ll 1$) is linearly proportional to ω_g and hence

[1] Recall that ω_g is a negative quantity.

linearly proportional to B_0. Since ω_g is negative for electrons, we again see that τ is positive (clockwise rotation) in the case of parallel propagation.

In the other special case, propagation being perpendicular to $\mathbf{B}_0$, that is, along the y axis, we have $\theta = \pi/2$, and Eqs. (151) reduce to

$$E_{0x}\left(1 - \frac{v^2}{c^2}\frac{\epsilon_{xx}}{\epsilon_0}\right) - E_{0y}\left(\frac{v^2}{c^2}\frac{\epsilon_{xy}}{\epsilon_0}\right) = 0 \tag{179}$$

$$E_{0x}\left(-\frac{v^2}{c^2}\frac{\epsilon_{yx}}{\epsilon_0}\right) + E_{0y}\left(-\frac{v^2}{c^2}\frac{\epsilon_{yy}}{\epsilon_0}\right) = 0 \tag{180}$$

$$E_{0z}\left(1 - \frac{v^2}{c^2}\frac{\epsilon_{zz}}{\epsilon_0}\right) = 0 \tag{181}$$

When in accord with Eq. (159) we choose

$$\frac{v^2}{c^2} = \frac{\epsilon_0}{\epsilon_{zz}} \tag{182}$$

then from Eqs. (179), (180), and (181) it follows that E_{0x} and E_{0y} are identically zero, and the only surviving component of the electric vector is E_{0z}. Thus we see that one of the two waves traveling in the y direction is a linearly polarized TEM wave whose electric vector is parallel to $\mathbf{B}_0$ and has the form

$$\mathbf{E}' = \mathbf{e}_z A e^{ik'_{\pi/2}y} \tag{183}$$

where A is an arbitrary constant. Since the propagation constant $k'_{\pi/2}$ as given by Eq. (161) is independent of B_0 and equal to the propagation constant of a wave in an isotropic plasma, this TEM wave (the ordinary wave) is independent of B_0 in its propagation properties and behaves as though it were a TEM wave in an isotropic plasma.

To obtain the extraordinary wave propagating perpendicular to $\mathbf{B}_0$, the other possible value of v^2/c^2 as given by Eq. (160) is used. That is,

$$\frac{v^2}{c^2} = \frac{\epsilon_{xx}/\epsilon_0}{(\epsilon_{xx}/\epsilon_0)^2 + (\epsilon_{xy}/\epsilon_0)^2} \tag{184}$$

is substituted into Eqs. (179), (180), and (181). Thus it is found that

E_{0z} vanishes identically and that

$$\frac{E_{0x}}{E_{0y}} = -\frac{\epsilon_{yy}}{\epsilon_{yx}} = i\frac{1 - X - Y^2}{XY} \tag{185}$$

Therefore the electric vector of this extraordinary wave has the form

$$\mathbf{E}'' = \left(i\mathbf{e}_x \frac{1 - X - Y^2}{XY} + \mathbf{e}_y\right) Ce^{ik''_{\pi/2}y} \tag{186}$$

where C is an arbitrary constant. The magnetic vector $\mathbf{H}''$ is obtained by substituting $\mathbf{E}''$ into the first of Eqs. (166). Thus

$$\mathbf{H}'' = -i\mathbf{e}_z \frac{k''_{\pi/2}}{\omega\mu_0} \frac{1 - X - Y^2}{XY} Ce^{ik''_{\pi/2}y} \tag{187}$$

From expressions (186) and (187), we see that the extraordinary wave traveling perpendicular to $\mathbf{B}_0$ is an $\mathbf{E}$ wave (TM wave) with its magnetic vector parallel to $\mathbf{B}_0$.

For propagation in an arbitrary direction θ, it follows from Eqs. (151) that the ratio ρ of the electric vector components perpendicular to $\mathbf{n}$ is given by[1]

$$\rho' = \frac{E'_x}{E'_\theta} = -\frac{i}{Y_L}\left[\frac{1}{2}\frac{Y_T{}^2}{1 - X} - \sqrt{\frac{1}{4}\frac{Y_T{}^4}{(1 - X)^2} + Y_L{}^2}\right] \tag{188}$$

for the ordinary wave whose propagation constant is k'_θ and by

$$\rho'' = \frac{E''_x}{E''_\theta} = -\frac{i}{Y_L}\left[\frac{1}{2}\frac{Y_T{}^2}{1 - X} + \sqrt{\frac{1}{4}\frac{Y_T{}^4}{(1 - X)^2} + Y_L{}^2}\right] \tag{189}$$

for the extraordinary wave whose propagation constant is k''_θ. Here E_θ is the component of $\mathbf{E}$ in the direction of the unit vector $\mathbf{e}_\theta$, which is defined by $\mathbf{e}_x \times \mathbf{e}_\theta = \mathbf{n}$. That is, $E_\theta = -E_z \sin\theta + E_y \cos\theta$. The ratio E_x/E_θ is a measure of the polarization of the part of $\mathbf{E}$ that is transverse to the direction of propagation $\mathbf{n}$ and is sometimes referred to as the polarization factor. The projection of the tip of $\mathbf{E}$ on a plane

[1] Without loss of generality, we still take $\mathbf{n}$ to lie in the zy plane.

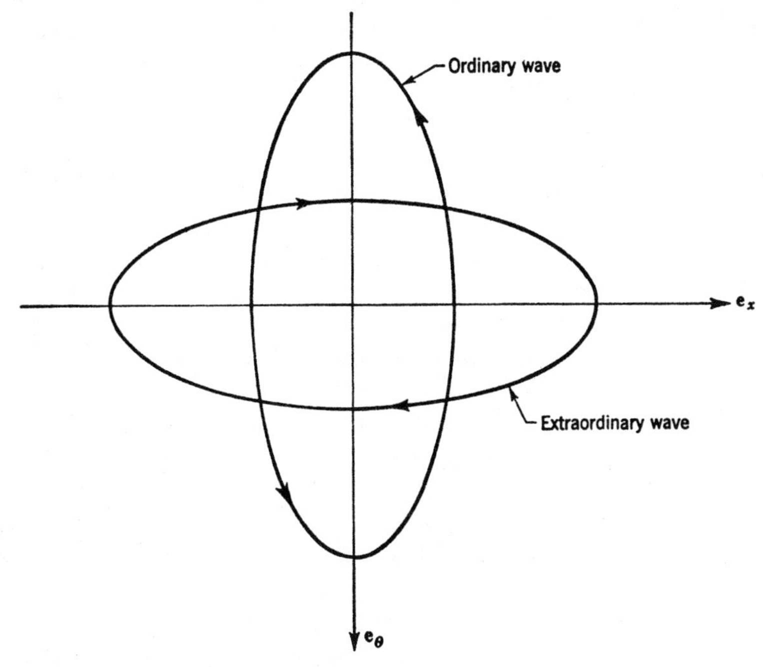

Fig. 6.3 ***Polarization ellipses of ordinary and extraordinary waves traveling into the plane of the paper. Ordinary wave is counterclockwise. Extraordinary wave is clockwise.***

transverse to **n** sweeps out an ellipse and, accordingly, the wave is said to be elliptically polarized. We note that $\rho'\rho'' = 1$ and consequently the ordinary and extraordinary waves are oppositely polarized. In the case of the ordinary wave the sense of polarization is counterclockwise and in the case of the extraordinary wave it is clockwise. See Fig. 6.3.

6.7 Antenna Radiation in Isotropic Plasma

So far we have been concerned with only the plane wave solutions of Maxwell's equations for a homogeneous plasma medium. Now, as a generalization to a case that involves spherical waves, we consider the far-zone radiation field of a primary source in an unbounded plasma. For simplicity, the primary source is taken to be a thin, center-driven, straight-wire antenna of length $2l$, and the ambient plasma is assumed to be homogeneous and isotropic. The antenna is driven monochromatically at an angular frequency ω and the time-average power fed into its input terminals is P_i. The problem is to find for fixed P_i and ω the far-zone radiation field of the antenna as a function of $X(=\omega_p^2/\omega^2)$.

Actually the basic part of the calculation has already been made in Chap. 3. Indeed, all we are required to do is to replace ϵ by $\epsilon_0(1-X)$ and k by $(\omega/c)\sqrt{1-X}$ in expressions (17), (18), and (19) of Sec. 3.2. However, since these expressions are valid only in the far zone, we must be careful not to violate the condition $(\omega/c)\sqrt{1-X}\,r \gg 1$. Clearly this condition can be met for the range $0 \leq X < 1$ by making r, the distance from the center of the antenna to the observation point, sufficiently large; but for $X = 1$ (plasma resonance) the condition is violated. Moreover, at $X = 1$ we have cutoff, i.e., no wave propagation can occur, and the power fed into the antenna goes into heating the plasma.

As in Sec. 3.2, we place the antenna along the z axis of a cartesian coordinate system, with one end of the antenna at $z = -l$ and the other end at $z = l$. With respect to the concentric spherical coordinate system (r,θ,ϕ) shown in Fig. 3.1, we see that the far-zone field components of the antenna immersed in a homogeneous isotropic plasma medium are

$$E_\theta = \frac{1}{\sqrt{1-X}}\sqrt{\frac{\mu_0}{\epsilon_0}}\,H_\phi \tag{190}$$

$$H_\phi = -\frac{ie^{i(\omega/c)\sqrt{1-X}r}}{2\pi r}\,I_0(X)F(\theta,X) \tag{191}$$

and the radial component of the time-average Poynting vector in the far zone is

$$S_r = \frac{1}{\sqrt{1-X}}\sqrt{\frac{\mu_0}{\epsilon_0}}\,\frac{I_0^2(X)}{8\pi^2 r^2}\,F^2(\theta,X) \tag{192}$$

This follows from Eqs. (17), (18), and (19) of Chap. 3 when ϵ is replaced by $\epsilon_0(1 - X)$ and k is replaced by $(\omega/c)\sqrt{1 - X}$. The radiation pattern $F(\theta,X)$ of the antenna is given by

$$F(\theta,X) \equiv \frac{\cos[(\omega/c)\sqrt{1 - X}\, l \cos\theta] - \cos[(\omega/c)\sqrt{1 - X}\, l]}{\sin\theta} \tag{193}$$

To find how I_0, the magnitude of the current at the driving point, depends on the time-average real power P_i fed into the antenna's input terminals and on the parameter $X(= \omega_p^2/\omega^2)$ which completely describes the plasma medium into which the antenna radiates, we note that since the plasma is assumed to be lossless, the time-average power P radiated by the antenna must be equal to P_i. Substituting expression (192) into the definition

$$P = \int_0^{2\pi} \int_0^{\pi} S_r r^2 \sin\theta \, d\theta \, d\phi \tag{194}$$

and equating P to P_i, we find that I_0 is related to P_i as follows:

$$P_i = \frac{1}{\sqrt{1 - X}} \sqrt{\frac{\mu_0}{\epsilon_0}} \frac{I_0^2(X)}{4\pi} \int_0^{\pi} F^2(\theta,X) \sin\theta \, d\theta \tag{195}$$

More conveniently, we write this relation in the form

$$P_i = \tfrac{1}{2} I_0^2(X) R_{rad}(X) \tag{196}$$

where the new parameter R_{rad}, the so-called radiation resistance of the antenna, has the representation

$$R_{rad}(X) = \frac{1}{\sqrt{1 - X}} \sqrt{\frac{\mu_0}{\epsilon_0}} \frac{1}{2\pi} \int_0^{\pi} F^2(\theta,X) \sin\theta \, d\theta \tag{197}$$

By substituting expression (193) into the integral and performing the operations that led to Eq. (101) of Chap. 3, we obtain

$$R_{rad}(X) = \frac{1}{\sqrt{1 - X}} \sqrt{\frac{\mu_0}{\epsilon_0}} \frac{1}{2\pi} \left[C + \ln 2\alpha - \operatorname{Ci} 2\alpha + \frac{\sin 2\alpha}{2} (\operatorname{Si} 4\alpha - 2\operatorname{Si} 2\alpha) + \frac{\cos 2\alpha}{2} (C + \ln\alpha + \operatorname{Ci} 4\alpha - 2\operatorname{Ci} 2\alpha) \right] \tag{198}$$

where $C(= 0.5722)$ is Euler's constant and $\alpha \equiv (\omega/c)\sqrt{1 - X}\,l$. Thus we see from Eq. (196) that I_0 depends on P_i and R_{rad} as follows,

$$I_0(X) = \sqrt{\frac{2P_i}{R_{rad}(X)}} \tag{199}$$

and from expression (198) that $R_{rad}(X)$ can be calculated for any X in the range $0 \leq X < 1$.

In view of relation (199) the far-zone field expressions (190) and (191) can be written as

$$E_\theta = -\frac{i}{\sqrt{1-X}}\sqrt{\frac{\mu_0}{\epsilon_0}}\frac{e^{i(\omega/c)\sqrt{1-X}r}}{2\pi r}\sqrt{\frac{2P_i}{R_{rad}(X)}}\,F(\theta,X) \tag{200}$$

$$H_\phi = -\frac{ie^{i(\omega/c)\sqrt{1-X}r}}{2\pi r}\sqrt{\frac{2P_i}{R_{rad}(X)}}\,F(\theta,X) \tag{201}$$

These are the desired forms because they show how the far-zone fields E_θ, H_ϕ depend on X. In the special case where $X = 0$ they reduce, as they should, to the conventional expressions for the far-zone fields of a straight-wire antenna in vacuum. In the other special case where P_i, ω, and l are fixed and X is made to approach unity, we find that

$$F(\theta,X) \to \frac{1}{2}\left(\frac{\omega}{c}\right)^2 l^2(1 - X)\sin\theta \tag{202}$$

$$R_{rad}(X) \to \frac{1}{6\pi}\sqrt{\frac{\mu_0}{\epsilon_0}}\left(\frac{\omega}{c}\right)^4 l^4(1 - X)^{3/2} \tag{203}$$

Consequently, as $X \to 1$, expressions (200) and (201) reduce to

$$E_\theta \sim \sqrt{\frac{\mu_0}{\epsilon_0}}\frac{e^{i(\omega/c)\sqrt{1-X}r}}{r}\sqrt{P_i}\,\frac{\sin\theta}{(1 - X)^{1/4}} \tag{204}$$

$$H_\phi \sim \frac{e^{i(\omega/c)\sqrt{1-X}r}}{r}\sqrt{P_i}\,(1 - X)^{1/4}\sin\theta \tag{205}$$

This shows that as $X \to 1$, the antenna's radiation pattern approaches the radiation pattern of a Hertzian dipole. It also shows that the wave impedance Z, which is given by

$$Z = \frac{E_\theta}{H_\phi} = \sqrt{\frac{\mu_0}{\epsilon_0}} \frac{1}{\sqrt{1 - X}} \tag{206}$$

increases without bound as $X \to 1$.

6.8 Dipole Radiation in Anisotropic Plasma

As was shown in Chap. 2, the radiation field of a monochromatic source in an unbounded homogeneous isotropic medium can be calculated by either the method of potentials or the method of the dyadic Green's function. As long as the medium is homogeneous and isotropic these two methods are equally convenient. However, in the case where the surrounding medium is anisotropic, the method of potentials[1] leads to difficulties in the early stages of the calculation and the Green's function method becomes the more fruitful of the two. Indeed, Bunkin,[2] Kogelnik,[3] and Kuehl[4] used the Green's function method with considerable success to analyze various aspects of the problem of a primary source in an anisotropic medium. Recalling some of their results, we shall now show how one proceeds in the Green's function method to find the radiation field of a dipole immersed in an unbounded homogeneous anisotropic plasma.

The electric field **E** of a monochromatic source **J** immersed in an

[1] A. Nisbet, Electromagnetic Potentials in a Heterogeneous Non-Conducting Medium, *Proc. Royal Soc. (London)*, (4) **240:** 375–381 (1957).

[2] F. V. Bunkin, On Radiation in Anisotropic Media, *J. Exptl. Theoret. Phys., U.S.S.R.*, **32:** 338–346 (1957); also *Soviet Physics JETP*, **5:** 277–283 (1957).

[3] H. Kogelnik, The Radiation Resistance of an Elementary Dipole in Anisotropic Plasmas, *Proc. Fourth Intern. Conf. on Ionization Phen. in Gases* (Uppsala, 1959), pp. 721–725, North Holland Publishing Company, Amsterdam, 1960. Also *J. Res. Natl. Bur. Std.*, **64D** (5): 515–523 (1960).

[4] H. Kuehl, Radiation from an Electric Dipole in an Anisotropic Cold Plasma, *Caltech Antenna Lab. Rept.* 24, October, 1960; also *Phys. Fluids*, **5:** 1095–1103 (1962).

unbounded anisotropic plasma medium must satisfy

$$\nabla \times \nabla \times \mathbf{E} - \omega^2\mu_0\boldsymbol{\varepsilon} \cdot \mathbf{E} = i\omega\mu_0\mathbf{J} \tag{207}$$

Moreover, $\mathbf{E}$ must have the form of a wave traveling away from the source. Hence, we are required to find the particular integral of Eq. (207) that satisfies the radiation condition.

By virtue of the linearity of Eq. (207), the desired solution may be expressed in the form

$$\mathbf{E}(\mathbf{r}) = i\omega\mu_0 \int \boldsymbol{\Gamma}(\mathbf{r},\mathbf{r}') \cdot \mathbf{J}(\mathbf{r}')dV' \tag{208}$$

where the integration extends throughout the region of finite extent occupied by the current. If this form is to be the solution of Eq. (207), the dyadic Green's function $\boldsymbol{\Gamma}(\mathbf{r},\mathbf{r}')$ must satisfy

$$\nabla \times \nabla \times \boldsymbol{\Gamma}(\mathbf{r},\mathbf{r}') - \omega^2\mu_0\boldsymbol{\varepsilon} \cdot \boldsymbol{\Gamma}(\mathbf{r},\mathbf{r}') = \mathbf{u}\delta(\mathbf{r} - \mathbf{r}') \tag{209}$$

or

$$\nabla\nabla \cdot \boldsymbol{\Gamma}(\mathbf{r},\mathbf{r}') - \nabla^2\boldsymbol{\Gamma}(\mathbf{r},\mathbf{r}') - \omega^2\mu_0\boldsymbol{\varepsilon} \cdot \boldsymbol{\Gamma}(\mathbf{r},\mathbf{r}') = \mathbf{u}\delta(\mathbf{r} - \mathbf{r}') \tag{210}$$

where $\mathbf{u}$ is the unit dyadic and $\delta(\mathbf{r} - \mathbf{r}')$ is the three-dimensional Dirac delta function.

To facilitate the construction of the dyadic Green's function, we express it as a Fourier integral. That is, we write

$$\boldsymbol{\Gamma}(\mathbf{r},\mathbf{r}') = \frac{1}{8\pi^3} \int_{-\infty}^{\infty} \boldsymbol{\Lambda}(\mathbf{k})e^{i\mathbf{k}\cdot(\mathbf{r}-\mathbf{r}')}\, d\mathbf{k} \tag{211}$$

and by so doing transform the problem of finding $\boldsymbol{\Gamma}$ into one of first finding the dyadic function $\boldsymbol{\Lambda}(\mathbf{k})$ and then evaluating the integral in $\mathbf{k}$ space. Substituting expression (211) into Eq. (210) and recalling the integral representation

$$\delta(\mathbf{r} - \mathbf{r}') = \frac{1}{8\pi^3} \int_{-\infty}^{\infty} e^{i\mathbf{k}\cdot(\mathbf{r}-\mathbf{r}')}\, d\mathbf{k} \tag{212}$$

we see that $\boldsymbol{\Lambda}(\mathbf{k})$ is determined by

$$\mathbf{V}(\mathbf{k}) \cdot \boldsymbol{\Lambda}(\mathbf{k}) = \mathbf{u} \tag{213}$$

where

$$\mathbf{V}(\mathbf{k}) = -\mathbf{kk} + k^2\mathbf{u} - \omega^2\mu_0\boldsymbol{\varepsilon} \tag{214}$$

With the aid of the theory of matrices, Eq. (213) yields for $\mathbf{\Lambda}(\mathbf{k})$ the expression

$$\mathbf{\Lambda}(\mathbf{k}) = \frac{\text{adj } \mathbf{V}(\mathbf{k})}{\det \mathbf{V}(\mathbf{k})} \tag{215}$$

Here det $\mathbf{V}(\mathbf{k})$ stands for the determinant of the matrix of $\mathbf{V}(\mathbf{k})$ and adj $\mathbf{V}(\mathbf{k})$ represents the dyadic whose matrix is the adjoint of the matrix of $\mathbf{V}(\mathbf{k})$.[1] It therefore follows from Eqs. (211) and (215) that the integral form of the dyadic Green's function is

$$\mathbf{\Gamma}(\mathbf{r},\mathbf{r}') = \frac{1}{8\pi^3}\int_{-\infty}^{\infty} \frac{\text{adj } \mathbf{V}(\mathbf{k})}{\det \mathbf{V}(\mathbf{k})} e^{i\mathbf{k}\cdot(\mathbf{r}-\mathbf{r}')}\, d\mathbf{k} \tag{216}$$

This form obeys the radiation condition and hence constitutes the only solution of Eq. (211) that leads to a physically acceptable result.

Since the source of radiation in the present instance is an oscillating electric dipole, we write the current distribution as

$$\mathbf{J}(\mathbf{r}') = -i\omega\mathbf{p}\delta(\mathbf{r}') \tag{217}$$

where $\mathbf{p}$ denotes the electric dipole moment. Substituting the current (217) and the Green's function (216) into the form (208), we obtain the integral representation

$$\mathbf{E}(\mathbf{r}) = \frac{\omega^2\mu_0}{8\pi^3}\int_{-\infty}^{\infty} \frac{[\text{adj } \mathbf{V}(\mathbf{k})]\cdot\mathbf{p}}{\det \mathbf{V}(\mathbf{k})} e^{i\mathbf{k}\cdot\mathbf{r}}\, d\mathbf{k} \tag{218}$$

which is the desired expression for the electric field $\mathbf{E}$ of the dipole $\mathbf{p}$. Thus we see that in the Green's function method the problem of calculating the field of a dipole in a homogeneous anisotropic medium splits into an algebraic part, which consists in finding the adjoint and the determinant of the matrix components of the dyadic $\mathbf{V}(\mathbf{k})$, and into

[1] See, for example, H. Margenau and G. M. Murphy, "The Mathematics of Physics and Chemistry," p. 295, D. Van Nostrand Company, Inc., Princeton, N.J., 1943.

an analytic part, which requires the evaluation of the integral in expression (218).

According to Kuehl, when the dipole oscillates at a high frequency, i.e., when $X = \omega_p^2/\omega^2 \ll 1$ and $Y^2 = \omega_g^2/\omega^2 \ll 1$, the dipole's far-zone electric field in the spherical coordinates r, θ, ϕ is given by

$$\mathbf{E} = -\left(\frac{\omega}{c}\right)^2 p_z \sin\theta \frac{e^{i(\omega/c)(1-X/2)r}}{4\pi\epsilon_0 r} (\mathbf{e}_\theta \cos\beta r - \mathbf{e}_\phi \sin\beta r) \tag{219}$$

for a z-directed dipole of moment p_z parallel to the biasing field $\mathbf{B}_0$, and by

$$\mathbf{E} = \left(\frac{\omega}{c}\right)^2 p_x \sqrt{1 - \sin^2\theta\cos^2\phi}\, \frac{e^{i(\omega/c)(1-X/2)r}}{4\pi\epsilon_0 r} [\mathbf{e}_\theta \cos(\beta r + \alpha) - \mathbf{e}_\phi \sin(\beta r + \alpha)] \tag{220}$$

for an x-directed dipole of moment p_x perpendicular to the biasing field $\mathbf{B}_0$. Here $\beta = k(\frac{1}{2})XY\cos\theta$ and $\alpha = \tan^{-1}(\tan\phi/\cos\theta)$. Comparing these expressions with the corresponding ones for a dipole in an isotropic plasma, we see that in the case of high frequencies the anisotropy does not change the amplitude $\sqrt{\mathbf{E}\cdot\mathbf{E}^*}$ of the radiated field $\mathbf{E}$ but does change its state of polarization: it causes the field to undergo Faraday rotation.

6.9 Reciprocity

Let $\mathbf{E}_1$, $\mathbf{H}_1$ be the electromagnetic field radiated by a current $\mathbf{J}_1$ occupying a finite volume V_1 and let $\mathbf{E}_2$, $\mathbf{H}_2$ be the electromagnetic field radiated by a current $\mathbf{J}_2$ occupying another finite volume V_2. The two source currents oscillate monochromatically at the same frequency and the medium occupying the space V_3 outside of V_1 and V_2 is anisotropic and may be inhomogeneous.

Clearly $\mathbf{E}_1$, $\mathbf{H}_1$ are related to $\mathbf{J}_1$ and $\mathbf{E}_2$, $\mathbf{H}_2$ are related to $\mathbf{J}_2$ by the equations

$$\nabla \times \mathbf{H}_1 = \mathbf{J}_1 - i\omega\boldsymbol{\varepsilon}\cdot\mathbf{E}_1 \qquad \nabla \times \mathbf{H}_2 = \mathbf{J}_2 - i\omega\boldsymbol{\varepsilon}\cdot\mathbf{E}_2 \tag{221}$$

Multiplying the first one by $\mathbf{E}_2$ and the second one by $\mathbf{E}_1$, and then subtracting the resulting equations, we get

$$\mathbf{E}_2 \cdot \nabla \times \mathbf{H}_1 - \mathbf{E}_1 \cdot \nabla \times \mathbf{H}_2 = \mathbf{E}_2 \cdot \mathbf{J}_1 - \mathbf{E}_1 \cdot \mathbf{J}_2 - i\omega \mathbf{E}_2 \cdot \boldsymbol{\varepsilon} \cdot \mathbf{E}_1 + i\omega \mathbf{E}_1 \cdot \boldsymbol{\varepsilon} \cdot \mathbf{E}_2 \tag{222}$$

With the aid of $\nabla \times \mathbf{E}_1 = i\omega\mu_0\mathbf{H}_1$ and $\nabla \times \mathbf{E}_2 = i\omega\mu_0\mathbf{H}_2$ we write the left side of Eq. (222) as a divergence and thus obtain

$$\nabla \cdot (\mathbf{E}_1 \times \mathbf{H}_2 - \mathbf{E}_2 \times \mathbf{H}_1) = \mathbf{E}_2 \cdot \mathbf{J}_1 - \mathbf{E}_1 \cdot \mathbf{J}_2 - i\omega \mathbf{E}_2 \cdot \boldsymbol{\varepsilon} \cdot \mathbf{E}_1 + i\omega \mathbf{E}_1 \cdot \boldsymbol{\varepsilon} \cdot \mathbf{E}_2 \tag{223}$$

Integrating this relation throughout all space and converting the left side of the resulting equation to a surface integral which vanishes by virtue of the behavior of the fields over the sphere at infinity, we are led to the expression

$$\int_{V_1} \mathbf{E}_2 \cdot \mathbf{J}_1 \, dV = \int_{V_2} \mathbf{E}_1 \cdot \mathbf{J}_2 \, dV + U \tag{224}$$

where

$$U = i\omega \int_{V_3} (\mathbf{E}_2 \cdot \boldsymbol{\varepsilon} \cdot \mathbf{E}_1 - \mathbf{E}_1 \cdot \boldsymbol{\varepsilon} \cdot \mathbf{E}_2) dV \tag{225}$$

When U is zero, Eq. (224) yields the relation

$$\int_{V_1} \mathbf{E}_2 \cdot \mathbf{J}_1 \, dV = \int_{V_2} \mathbf{E}_1 \cdot \mathbf{J}_2 \, dV \tag{226}$$

which defines what we usually mean by reciprocity.[1] That is, two monochromatic sources are said to be reciprocal when the source cur-

[1] The reciprocity theorem for electromagnetic waves is a generalization of Rayleigh's reciprocity theorem for sound waves (see Lord Rayleigh, "Theory of Sound," 2d ed., vol. II, pp. 145–148, Dover Publications, Inc., New York, 1945) and stems from the work of Lorentz [see H. A. Lorentz, *Amsterdammer Akademie van Wetenschappen*, 4: 176 (1895–1896)]. For a detailed discussion see P. Poincelot, "Précis d'électromagnétisme théorique," chap. 18, Dunod, Paris, 1963.

rents and their radiated electric fields satisfy relation (226) or, equivalently, when the quantity U vanishes. Clearly, U vanishes when the dielectric constant of the medium is symmetric ($\epsilon_{ik} = \epsilon_{ki}$). However, for a magnetically biased plasma the dielectric constant is hermitian and hence U does not necessarily vanish. This means that in the case of an anisotropic plasma reciprocity does not necessarily hold.

Nevertheless, the concept of reciprocity can be generalized, at least formally, to include the case of an anisotropic plasma.[1] Such a generalization is based on the fact that the dielectric tensor of a magnetically biased plasma is symmetrical under a reversal of the biasing magnetostatic field, i.e.,

$$\epsilon_{ik}(\mathbf{B}_0) = \epsilon_{ki}(-\mathbf{B}_0) \tag{227}$$

or

$$\boldsymbol{\varepsilon}(\mathbf{B}_0) = \tilde{\boldsymbol{\varepsilon}}(-\mathbf{B}_0) \tag{228}$$

where the tilde indicates the transposed dyadic. When the biasing field is $\mathbf{B}_0$, we have for the fields produced by $\mathbf{J}_1$ the Maxwell equation

$$\nabla \times \mathbf{H}_1(\mathbf{B}_0) = \mathbf{J}_1 - i\omega\boldsymbol{\varepsilon}(\mathbf{B}_0) \cdot \mathbf{E}_1(\mathbf{B}_0) \tag{229}$$

Moreover, when the biasing field is $-\mathbf{B}_0$, we have for the fields produced by $\mathbf{J}_2$ the Maxwell equation

$$\nabla \times \mathbf{H}_2(-\mathbf{B}_0) = \mathbf{J}_2 - i\omega\boldsymbol{\varepsilon}(-\mathbf{B}_0) \cdot \mathbf{E}_2(-\mathbf{B}_0) \tag{230}$$

which, in view of the symmetry relation (228), assumes the form

$$\nabla \times \mathbf{H}_2(-\mathbf{B}_0) = \mathbf{J}_2 - i\omega\tilde{\boldsymbol{\varepsilon}}(\mathbf{B}_0) \cdot \mathbf{E}_2(-\mathbf{B}_0) \tag{231}$$

Proceeding as before, we find from Eqs. (229) and (231) the relation

$$\int_{V_1} \mathbf{E}_2(-\mathbf{B}_0) \cdot \mathbf{J}_1 \, dV = \int_{V_2} \mathbf{E}_1(\mathbf{B}_0) \cdot \mathbf{J}_2 \, dV \tag{232}$$

[1] For application to ionospheric propagation see K. G. Budden, A Reciprocity Theorem on the Propagation of Radio Waves via the Ionosphere, *Proc. Cambridge Phil. Soc.*, **50:** 604 (1954).

which is the desired generalization of the reciprocity theorem to the case of an anisotropic plasma.[1] If $\mathbf{J}_2$ is such that $\mathbf{E}_2(-\mathbf{B}_0) = \mathbf{E}_2(\mathbf{B}_0)$, or if $\mathbf{J}_1$ is such that $\mathbf{E}_1(\mathbf{B}_0) = \mathbf{E}_1(-\mathbf{B}_0)$, this relation reduces to the usual reciprocal relation (226).

[1] Reciprocity and reversibility are not unrelated properties. If the current density $\mathbf{J}$ transforms into $-\mathbf{J}'$ when t is replaced by $-t'$, the Maxwell equations can be made invariant under time reversal by replacing $\mathbf{D}$ by $\mathbf{D}'$, $\mathbf{H}$ by $-\mathbf{H}''$, $\mathbf{B}$ by $-\mathbf{B}'$, and $\mathbf{E}$ by $\mathbf{E}'$. However, in a lossy medium the presence of a conduction current term $\sigma\mathbf{E}$ makes it impossible for the Maxwell equations to be invariant under time reversal.

The Doppler effect 7

If a source of monochromatic radiation is in motion relative to an observer, the observed frequency of radiation will increase as the source and observer approach each other and will decrease as they get farther apart. This principle, enunciated by Christian Doppler[1] in 1843, is called the "Doppler principle" or the "Doppler effect."

Basically the Doppler effect is a consequence of the covariance of Maxwell's equations under the Lorentz transformation. For the usual case where the source and observer are in free space, the exact relativistic formulation of the Doppler effect is well known. But in the presence of material media the Doppler effect is more intricate and involves questions which as yet have not been completely settled.

In this chapter the problem of calculating the Doppler effect in material media is discussed. It is shown that for homogeneous media the calculation can be made by using the principle of phase invariance, whereas for inhomogeneous media a more elementary point of departure is required.

[1] Ch. Doppler, Über das farbige Licht der Doppelsterne, *Abhandlungen der Koniglichen Bohmischen Gesellschaft der Wissenschaften*, 1843. See also E. N. Da C. Andrade, Doppler and the Doppler Effect, *Endeavor*, vol. 18, no. 69, January, 1959.

7.1 Covariance of Maxwell's Equations

According to the theory of relativity, the Maxwell equations must have the same form in all inertial frames of reference, i.e., they must be covariant under the Lorentz transformation.[1] This means that if we write the Maxwell equations in an inertial frame K and then by a proper Lorentz transformation pass from the coordinates x, y, z, t of K to the coordinates x', y', z', t' of another inertial frame K' which is moving at a uniform velocity with respect to K, the dependent functions, i.e., the four field vectors, the current density vector, and the charge density, must transform in such a way that the transformed equations have the same formal appearance as the original equations.

The Lorentz transformations can be considered a consequence of the postulate that the velocity of light in vacuum has the same value c in all frames of reference. To show this, we make the spatial origins of K and K' coincident at $t = t' = 0$ and introduce the convenient notation $x_1 = x$, $x_2 = y$, $x_3 = z$, $x_4 = ict$, $x'_1 = x'$, $x'_2 = y'$, $x'_3 = x'$, $x'_4 = ict'$. Then, in this notation, the postulate demands that the condition

$$x'_\mu x'_\mu = x_\mu x_\mu \tag{1}$$

be satisfied. Here and in analogous cases we suppress the summation sign and use the convention that repeated indices are summed from 1 to 4. This condition in turn leads to the requirement that the coordi-

[1] The covariance of the Maxwell equations under the Lorentz transformation was proved by Lorentz and Poincaré, and physically interpreted by Einstein. Their work, however, was intentionally restricted to the Maxwell equations of electron theory, i.e., to the so-called microscopic Maxwell-Lorentz equations, and said nothing of material media. The required generalization of the theory to the case of material media was finally worked out by Minkowski from the postulate that the macroscopic Maxwell equations are covariant under the Lorentz transformation. See, for example, W. Pauli, "Theory of Relativity," Pergamon Press, New York, 1958; A. Sommerfeld, "Electrodynamics," Academic Press Inc., New York, 1952; V. Fock, "The Theory of Space Time and Gravitation," Pergamon Press, New York, 1952; E. Whittaker, "A History of the Theories of Aether and Electricity," vol. II, Harper & Row, Publishers, Incorporated, New York, 1953; C. Møller, "The Theory of Relativity," Oxford University Press, Fair Lawn, N.J., 1952.

nates x'_μ and x_μ be related by the linear transformations

$$x'_\mu = a_{\mu\nu}x_\nu \qquad x_\nu = a_{\mu\nu}x'_\mu \tag{2}$$

whose coefficients $a_{\mu\nu}$ obey the side conditions

$$a_{\mu\nu}a_{\mu\lambda} = a_{\nu\mu}a_{\lambda\mu} = \delta_{\nu\lambda} = \begin{cases} 1 & \text{for} \quad \nu = \lambda \\ 0 & \text{for} \quad \nu \neq \lambda \end{cases} \tag{3}$$

These linear transformations constitute the complete Lorentz group of transformations. Since the determinant $|a_{\mu\nu}|$ may equal $+1$ or -1, this complete group splits naturally into the positive transformations for which $|a_{\mu\nu}| = 1$ and the negative transformations for which $|a_{\mu\nu}| = -1$. From these the positive transformations are selected because they include the identity transformation

$$x'_\mu = x_\mu \qquad (\mu = 1, 2, 3, 4) \tag{4}$$

The positive transformations, which can be thought of as a rotation in four-dimensional space or, equivalently, as six rotations in the x_1x_2, x_1x_3, x_1x_4, x_2x_3, x_2x_4, x_3x_4 planes, contain not only the proper Lorentz transformations but also extraneous transformations involving the reversal of two or four axes. Therefore, when these extraneous transformations are excluded, those that remain of the positive transformations constitute the proper Lorentz transformations.

Assuming that the coordinates undergo a proper Lorentz transformation, we define a 4-vector as a set of four quantities A_μ ($\mu = 1, 2, 3, 4$) that transform like the coordinates:

$$A'_\mu = a_{\mu\nu}A_\nu \tag{5}$$

Moreover, we define a 4-tensor $A_{\mu\nu}$ of rank 2 as a set of 4^2 quantities that obey the transformation law

$$A'_{\mu\nu} = a_{\mu\lambda}a_{\nu\eta}A_{\lambda\eta} \tag{6}$$

and a 4-tensor $A_{\mu\nu\lambda}$ of rank 3 as a set of 4^3 quantities that obey the transformation law

$$A'_{\mu\nu\lambda} = a_{\mu\alpha}a_{\nu\beta}a_{\lambda\gamma}A_{\alpha\beta\gamma} \tag{7}$$

In terms of the quantities $F_{\alpha\beta}$, $G_{\alpha\beta}$, J_α $(\alpha, \beta = 1, 2, 3, 4)$, whose values are given by

$$F_{\alpha\beta} = \begin{bmatrix} 0 & B_z & -B_y & -\frac{i}{c}E_x \\ -B_z & 0 & B_x & -\frac{i}{c}E_y \\ B_y & -B_x & 0 & -\frac{i}{c}E_z \\ \frac{i}{c}E_x & \frac{i}{c}E_y & \frac{i}{c}E_z & 0 \end{bmatrix} \tag{8}$$

$$G_{\alpha\beta} = \begin{bmatrix} 0 & H_z & -H_y & -icD_x \\ -H_z & 0 & H_x & -icD_y \\ H_y & -H_x & 0 & -icD_z \\ icD_x & icD_y & icD_z & 0 \end{bmatrix} \tag{9}$$

$$J_\alpha = \begin{bmatrix} J_x \\ J_y \\ J_z \\ ic\rho \end{bmatrix} \tag{10}$$

the two Maxwell equations

$$\nabla \cdot \mathbf{B} = 0 \qquad \nabla \times \mathbf{E} = -\frac{\partial}{\partial t}\mathbf{B} \tag{11}$$

become

$$\frac{\partial F_{\alpha\beta}}{\partial x_\gamma} + \frac{\partial F_{\beta\gamma}}{\partial x_\alpha} + \frac{\partial F_{\gamma\alpha}}{\partial x_\beta} = 0 \qquad (\alpha, \beta, \gamma = 1, 2, 3, 4) \tag{12}$$

and the other two Maxwell equations

$$\nabla \times \mathbf{H} - \frac{\partial}{\partial t}\mathbf{D} = \mathbf{J} \qquad \nabla \cdot \mathbf{D} = \rho \tag{13}$$

become

$$\frac{\partial G_{\alpha\beta}}{\partial x_\beta} = J_\alpha \qquad (\alpha = 1, 2, 3, 4) \tag{14}$$

From the postulate that Maxwell's equations are covariant under a proper Lorentz transformation of the coordinates, viz., that the four-dimensional forms (12) and (14) are covariant, it follows that $F_{\alpha\beta}$ and $G_{\alpha\beta}$ are 4-tensors of rank 2 and J_α is a 4-vector. This means that when the coordinates undergo a proper Lorentz transformation

$$x'_\mu = a_{\mu\nu}x_\nu \qquad (\mu = 1, 2, 3, 4) \tag{15}$$

the quantities J_α (the 4-current) transform like the coordinates:

$$J'_\mu = a_{\mu\alpha}J_\alpha \qquad (\mu = 1, 2, 3, 4) \tag{16}$$

and the field tensors $F_{\alpha\beta}$, $G_{\alpha\beta}$ transform like the product of the coordinates:

$$F'_{\mu\nu} = a_{\mu\alpha}a_{\nu\beta}F_{\alpha\beta} \qquad (\mu,\ \nu = 1, 2, 3, 4) \tag{17}$$

$$G'_{\mu\nu} = a_{\mu\alpha}a_{\nu\beta}G_{\alpha\beta} \qquad (\mu,\ \nu = 1, 2, 3, 4) \tag{18}$$

So far the only restrictions we have placed on the reference frames are that their spatial origins be coincident at $t = t' = 0$ and that their relative velocity $\mathbf{v}$ be uniform. Now we shall place an additional restriction on the reference frames, namely, that they have the same orientation. With the velocity and orientation specified, the coefficients $a_{\mu\nu}$ can be uniquely determined from Eqs. (2) and (3) and the condition $|a_{\mu\nu}| = 1$. One can show that if the two inertial frames K and K' have the same orientation, and if their relative velocity is $\mathbf{v}$, then the coefficients $a_{\mu\nu}$ are given by

$$a_{\mu\nu} = \begin{bmatrix} 1 + (\gamma - 1)\frac{v_x^2}{v^2} & (\gamma - 1)\frac{v_xv_y}{v^2} & (\gamma - 1)\frac{v_xv_z}{v^2} & i\gamma\frac{v_x}{c} \\ (\gamma - 1)\frac{v_yv_x}{v^2} & 1 + (\gamma - 1)\frac{v_y^2}{v^2} & (\gamma - 1)\frac{v_yv_z}{v^2} & i\gamma\frac{v_y}{c} \\ (\gamma - 1)\frac{v_zv_x}{v^2} & (\gamma - 1)\frac{v_zv_y}{v^2} & 1 + (\gamma - 1)\frac{v_z^2}{v^2} & i\gamma\frac{v_z}{c} \\ -i\gamma\frac{v_x}{c} & -i\gamma\frac{v_y}{c} & -i\gamma\frac{v_z}{c} & \gamma \end{bmatrix} \tag{19}$$

Using these values of the coefficients and expressing the results in three-dimensional form, we find that the transformation law (15) for the position 4-vector x_μ, which can be written as $(\mathbf{r},ict)$, becomes

$$\mathbf{r}' = \mathbf{r} - \gamma\mathbf{v}t + (\gamma - 1)\frac{(\mathbf{r}\cdot\mathbf{v})}{v^2}\mathbf{v} \tag{20}$$

$$t' = \gamma\left(t - \frac{\mathbf{r}\cdot\mathbf{v}}{c^2}\right) \tag{21}$$

where

$$\gamma = \frac{1}{\sqrt{1-\beta^2}} \qquad \beta = \frac{v}{c} \qquad \mathbf{r} = \mathbf{e}_x x + \mathbf{e}_y y + \mathbf{e}_z z$$

and that the transformation law (16) for the 4-vector $(\mathbf{J},ic\rho)$ assumes the form

$$\mathbf{J}' = \mathbf{J} - \gamma\mathbf{v}\rho + (\gamma - 1)\frac{\mathbf{J}\cdot\mathbf{v}}{v^2}\mathbf{v} \tag{22}$$

$$\rho' = \gamma\left(\rho - \frac{1}{c^2}\mathbf{J}\cdot\mathbf{v}\right) \tag{23}$$

Also, we find that the transformation law (17) leads to

$$\mathbf{E}' = \gamma(\mathbf{E} + \mathbf{v}\times\mathbf{B}) + (1-\gamma)\frac{\mathbf{E}\cdot\mathbf{v}}{v^2}\mathbf{v} \tag{24}$$

$$\mathbf{B}' = \gamma\left(\mathbf{B} - \frac{1}{c^2}\mathbf{v}\times\mathbf{E}\right) + (1-\gamma)\frac{\mathbf{B}\cdot\mathbf{v}}{v^2}\mathbf{v} \tag{25}$$

and that the transformation law (18) yields

$$\mathbf{D}' = \gamma\left(\mathbf{D} + \frac{1}{c^2}\mathbf{v}\times\mathbf{H}\right) + (1-\gamma)\frac{\mathbf{D}\cdot\mathbf{v}}{v^2}\mathbf{v} \tag{26}$$

$$\mathbf{H}' = \gamma(\mathbf{H} - \mathbf{v}\times\mathbf{D}) + (1-\gamma)\frac{\mathbf{H}\cdot\mathbf{v}}{v^2}\mathbf{v} \tag{27}$$

Clearly Eqs. (22) and (23) follow from Eqs. (20) and (21) by replacing $\mathbf{r}$ by $\mathbf{J}$ and ict by $ic\rho$. Also Eqs. (26) and (27) follow from Eqs. (24) and (25) by replacing $\mathbf{E}$ by $c\mathbf{D}$ and $\mathbf{B}$ by $\mathbf{H}/c$.

Thus we see that when the coordinates and time undergo the proper Lorentz transformations expressed by Eqs. (20) and (21), the Maxwell equations with respect to K, viz.,

$$\nabla \times \mathbf{H} = \mathbf{J} + \frac{\partial}{\partial t}\mathbf{D} \qquad \nabla \times \mathbf{E} = -\frac{\partial}{\partial t}\mathbf{B}$$
$$\nabla \cdot \mathbf{D} = \rho \qquad \nabla \cdot \mathbf{B} = 0 \tag{28}$$

transform into the Maxwell equations with respect to K', viz.,

$$\nabla' \times \mathbf{H}' = \mathbf{J}' + \frac{\partial}{\partial t'}\mathbf{D}' \qquad \nabla' \times \mathbf{E}' = -\frac{\partial}{\partial t'}\mathbf{B}'$$
$$\nabla' \cdot \mathbf{D}' = \rho' \qquad \nabla' \cdot \mathbf{B}' = 0 \tag{29}$$

provided the primed quantities are related to the unprimed quantities by relations (20) through (27).

7.2 Phase Invariance and Wave 4-Vector

If a reference frame K is at rest with respect to a homogeneous medium, the Maxwell equations in K admit solutions of the form

$$\mathbf{E}(\mathbf{r},t) = \operatorname{Re} \mathbf{E}_0 e^{i(\mathbf{k}\cdot\mathbf{r}-\omega t)} \tag{30}$$

$$\mathbf{B}(\mathbf{r},t) = \operatorname{Re} \mathbf{B}_0 e^{i(\mathbf{k}\cdot\mathbf{r}-\omega t)} \tag{31}$$

where $\mathbf{E}_0$ is a constant and $\mathbf{B}_0$, which is related to $\mathbf{E}_0$ by $\mathbf{B}_0 = (1/\omega)\mathbf{k} \times \mathbf{E}_0$, is likewise a constant. Expressions (30) and (31) represent in K the electric and magnetic vectors of a plane homogeneous wave of angular frequency ω and wave vector $\mathbf{k}$.

To see what form this plane wave takes in a reference frame K' moving at uniform velocity $\mathbf{v}$ with respect to K, we first substitute expressions (30) and (31) into the transformation law (24) and thus obtain the expression

$$\mathbf{E}'(\mathbf{r},t) = \operatorname{Re} \mathbf{E}'_0 e^{i(\mathbf{k}\cdot\mathbf{r}-\omega t)} \tag{32}$$

where $\mathbf{E}'_0$ is a constant given by

$$\mathbf{E}'_0 = \gamma(\mathbf{E}_0 + \mathbf{v} \times \mathbf{B}_0) + (1 - \gamma)\frac{\mathbf{E}_0 \cdot \mathbf{v}}{v^2}\mathbf{v} \tag{33}$$

Then we transform the coordinates $\mathbf{r}$ and t into the coordinates $\mathbf{r}'$ and t' of K' by means of the proper Lorentz transformation

$$\mathbf{r} = \mathbf{r}' + \gamma\mathbf{v}t' + (\gamma - 1)\frac{\mathbf{r}' \cdot \mathbf{v}}{v^2}\mathbf{v} \tag{34}$$

$$t = \gamma\left(t' + \frac{\mathbf{r}' \cdot \mathbf{v}}{c^2}\right) \tag{35}$$

Applying this transformation to expression (32), we see that the electric vector of the wave in K' takes the form

$$\mathbf{E}'(\mathbf{r}',t') = \operatorname{Re} \mathbf{E}'_0 e^{i(\mathbf{k}'\cdot\mathbf{r}' - \omega' t')} \tag{36}$$

where

$$\mathbf{k}' = \mathbf{k} - \gamma\frac{\omega}{c^2}\mathbf{v} + (\gamma - 1)\frac{\mathbf{k} \cdot \mathbf{v}}{v^2}\mathbf{v} \tag{37}$$

$$\omega' = \gamma(\omega - \mathbf{v} \cdot \mathbf{k}) \tag{38}$$

This shows that in going from K to K' the plane wave (30) is transformed into the plane wave (36).

By the manner in which $\mathbf{k}'$ and ω' appear in expression (36), we are led to the interpretation that $\mathbf{k}'$ is the wave vector of the wave in K' and ω' is its frequency. Accordingly, we regard relations (37) and (38) as the transformation laws for the wave vector and the frequency. Comparing these relations with Eqs. (20) and (21), we see that $\left(\mathbf{k}, i\frac{\omega}{c}\right)$ transforms like the 4-vector $(\mathbf{r}, ict)$. Hence

$$k_\mu = \begin{bmatrix} k_x \\ k_y \\ k_z \\ i\dfrac{\omega}{c} \end{bmatrix} \tag{39}$$

is a 4-vector. It is called the wave 4-vector.

The phase ϕ of the wave in K is defined by

$$\phi = \mathbf{k} \cdot \mathbf{r} - \omega t \tag{40}$$

and in terms of k_μ and x_μ it takes the form

$$\phi = k_\mu x_\mu \tag{41}$$

Since k_μ and x_μ are 4-vectors, it follows from Eq. (41) that ϕ is invariant.

What we have shown above is that the phase ϕ of a uniform plane wave in a homogeneous medium remains invariant under a proper Lorentz transformation of the coordinates. This invariance of the phase, sometimes referred to as the principle of phase invariance, applies not only to waves in vacuum but also to waves in homogeneous media, even if these homogeneous media be anisotropic and dispersive. However, in the case of inhomogeneous media the Maxwell equations do not admit uniform plane wave solutions and hence preclude the possibility of devising an invariant phase.[1]

7.3 Doppler Effect and Aberration

As in the previous section, we consider a plane monochromatic wave traveling in a homogeneous medium. We recall that if $\mathbf{k}$ and ω are respectively the wave vector and angular frequency of the wave in the reference frame K, which is at rest with respect to the medium, then the wave vector $\mathbf{k}'$ and the angular frequency ω' of the wave, as observed in a reference frame K' moving with uniform velocity $\mathbf{v}$ with respect to K, are given by

$$\mathbf{k}' = \mathbf{k} - \gamma \frac{\omega}{c^2} \mathbf{v} + (\gamma - 1) \frac{\mathbf{k} \cdot \mathbf{v}}{v^2} \mathbf{v} \tag{42}$$

$$\omega' = \gamma(\omega - \mathbf{v} \cdot \mathbf{k}) \tag{43}$$

[1] K. S. H. Lee and C. H. Papas, Doppler Effects in Inhomogeneous Anisotropic Ionized Gases, *J. Math. Phys.*, **42** (3): 189–199 (September, 1963).

where

$$\gamma = \frac{1}{\sqrt{1-\beta^2}} \qquad \beta^2 = \frac{\mathbf{v}\cdot\mathbf{v}}{c^2}$$

From Eq. (42) we can calculate the angle between the directions of $\mathbf{k}'$ and $\mathbf{k}$ and thus obtain the aberration of the wave vector due to the relative motion of the reference frames. Also, from (43) we can calculate the difference between ω' and ω, which gives the corresponding Doppler shift in frequency.

To derive the aberration formula, we note that the spatial axes of K and K' are similarly oriented, i.e., the x', y', z' axes are parallel respectively to the x, y, z axes, and we assume that $\mathbf{v}$ is parallel to the x axis and hence to the x' axis. Since $\mathbf{v} = \mathbf{e}_x v$, it follows from the scalar multiplication of Eq. (42) by the unit vectors $\mathbf{e}_x$ and $\mathbf{e}_y$ that

$$k' \cos\theta' = \gamma k \cos\theta - \gamma \frac{\omega}{c^2} v \tag{44}$$

$$k' \sin\theta' = k \sin\theta \tag{45}$$

where θ' is the angle between $\mathbf{k}'$ and $\mathbf{v}$, and θ is the angle between $\mathbf{k}$ and $\mathbf{v}$. Dividing Eq. (45) by Eq. (44) and using the relations $k = \omega/v_{\text{ph}}$, where v_{ph} is the phase velocity in K, $n = c/v_{\text{ph}}$, where n is the index of refraction in K, and $\beta = v/c$, we get the aberration formula

$$\tan\theta' = \frac{1}{\gamma}\frac{\sin\theta}{\cos\theta - \dfrac{\beta}{n}} = \frac{1}{\gamma}\frac{\tan\theta}{1 - \dfrac{\beta}{n}\sec\theta} \tag{46}$$

In vacuum, we have $n = 1$ and, accordingly, Eq. (46) reduces to the familiar relativistic formula for aberration.

The formula (43) for the Doppler effect can be written as

$$\omega' = \gamma(\omega - vk\cos\theta) = \gamma\omega(1 - \beta n \cos\theta) \tag{47}$$

where θ is the angle between the wave vector $\mathbf{k}$ and the relative velocity $\mathbf{v}$. From this equation we see that a wave of angular frequency ω in reference frame K appears to have a different frequency ω' when observed from the moving frame K'. The Doppler shift in frequency, viz., the quantity $\omega' - \omega$, is a maximum when $\theta = 0$ and is a minimum when $\theta = \pi/2$. In the latter case we have the relation

$$\omega' = \gamma\omega \tag{48}$$

which expresses the so-called "transverse Doppler effect."

7.4 Doppler Effect in Homogeneous Dispersive Media

We shall now apply the Doppler formula to the situation in which a monochromatic source and an observer are in a homogeneous dispersive medium. We shall limit the discussion to two cases: in one the source is fixed with respect to the medium and in the other the observer is fixed with respect to the medium. The observer is assumed to be in the far field of the source so that, to a good approximation, the waves incident upon the observer are plane.

In the case where the source is fixed with respect to the medium, we choose the reference frame K to be at rest with respect to the medium and the source, and the reference frame K' to be moving with the observer at velocity $\mathbf{v}$ with respect to K. Hence, from Eq. (47) we see that

$$\omega' = \gamma\omega[1 - \beta n(\omega) \cos \theta] \tag{49}$$

where ω is the source frequency in K, and ω' is the frequency observed in the moving frame K'. The index of refraction $n(\omega)$ is evaluated in K. Since $n(\omega) \geq 0$, it follows from expression (49) that when the observer is moving toward the source ($\theta = \pi$), ω' is greater than ω, as in a vacuum. However, when the observer is moving away from the source ($\theta = 0$), ω' is not necessarily less than ω. Under special circumstances (for example, when the medium is a nearly resonant plasma), $n(\omega)$ could be so small that ω' would be greater than ω, in contradistinction to the corresponding phenomenon in a vacuum, where ω' would necessarily have to be less than ω.

In the case where the observer is fixed with respect to the medium, we choose K to be at rest with respect to the medium and the observer, and K' to be moving with the source at velocity $\mathbf{v}$ with respect to K. Accordingly we again have

$$\omega' = \gamma\omega[1 - \beta n(\omega) \cos \theta] \tag{50}$$

but now ω' is the source frequency and ω the observed frequency. When $\theta = \pi$ the source is moving away from the observer, and when $\theta = 0$ it is moving toward the observer. Due to the dispersive nature of $n(\omega)$, expression (50) is not, in general, monotonic between ω' and ω. Therefore, a given value of ω' may yield more than one value of ω. This means that the radiation incident upon the observer may appear to have several spectral components even though the source is oscillating at a single frequency. This splitting of the emitted monochromatic radiation into several modes is called the complex Doppler effect. This effect has been studied by Frank[1] in connection with the problem of determining the radiation of an oscillating dipole moving through a refractive medium. If the medium were nondispersive, expression (50) would, of course, yield a monotonic relation between ω' and ω, and hence no complex Doppler modes would be generated.

As an illustrative example, let us examine the complex Doppler effect in the special instance where the medium is a homogeneous plasma. For such a medium, Eq. (50) becomes

$$\omega' = \gamma(\omega - \beta \sqrt{\omega^2 - \omega_p{}^2} \cos \theta) \tag{51}$$

where ω_p is the plasma frequency. A plot of ω' versus ω is shown in Fig. 7.1. The curve has two branches, one given by the solid line and the other by the broken line. The broken line represents Eq. (51) for $\theta = \pi$ (source receding from the observer), and the solid line represents Eq. (51) for $\theta = 0$ (source approaching the observer). The two branches join at point A, where $\omega = \omega_p$ and $\omega' = \gamma\omega_p$. The solid branch is a minimum at point B, where $\omega = \gamma\omega_p$ and $\omega' = \omega_p$. The asymptotes make with the axes an angle ψ which depends on the relative

[1] I. M. Frank, Doppler Effect in a Refractive Medium, *J. Phys. U.S.S.R.*, 7 (2): 49–67 (1943). See also, O. E. H. Rydbeck, *Chalmers Res. Rept.* 10, 1960.

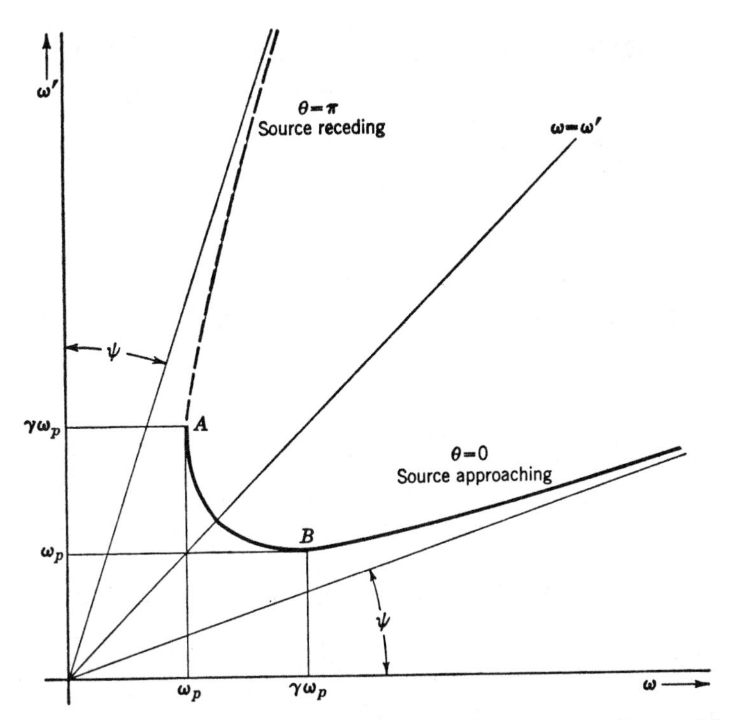

Fig. 7.1 ***A sketch of the source frequency ω' versus the observed frequency ω, in the case where the observer is at rest with respect to a homogeneous isotropic plasma medium and the source is moving through the medium at relative velocity βc.***

velocity v according to the relation $\tan \psi = \sqrt{1 - \beta}/\sqrt{1 + \beta}$. From the curve, we see that for a given value ω'_s of ω' greater than $\gamma\omega_p$, we get a single value ω_r of ω when the source is receding, and a single value ω_a of ω when the source is approaching. We also see that if ω'_s is less than $\gamma\omega_p$ but greater than ω_p, the wave due to the receding source is beyond cutoff, and the wave due to the approaching source splits into two, thus yielding two values of ω_a instead of only one. One of these two frequencies is always greater than ω'_s, while the other may be greater or less than ω'_s depending on how close ω'_s is to $\gamma\omega_p$. Finally, we note that if ω'_s is less than ω_p, even the wave due to the approaching source is beyond cutoff.

7.5 Index of Refraction of a Moving Homogeneous Medium

To compute the index of refraction of a homogeneous medium moving at velocity $\mathbf{v}$ with respect to a reference frame K, we choose a frame K' that is at rest with respect to the medium, and we assume that in K' there is a monochromatic plane wave having wave vector $\mathbf{k}'$ and frequency ω'. In K the wave is perceived as a plane wave of wave vector $\mathbf{k}$ and frequency ω. The index of refraction of the medium is defined by $n' = ck'/\omega'$ in K' and by $n = ck/\omega$ in K.

As a point of departure for the calculation, we use the transformations

$$\mathbf{k} = \mathbf{k}' + \gamma \frac{\omega'}{c^2} \mathbf{v} + (\gamma - 1) \frac{\mathbf{k}' \cdot \mathbf{v}}{v^2} \mathbf{v} \tag{52}$$

$$\omega = \gamma(\omega' + \mathbf{v} \cdot \mathbf{k}') \tag{53}$$

From Eq. (52) we find that k is given by

$$k = \left[k'^2 + 2\gamma^2 \frac{\omega'}{c^2} \mathbf{k}' \cdot \mathbf{v} + (\gamma^2 - 1) \frac{(\mathbf{k}' \cdot \mathbf{v})^2}{v^2} + \gamma^2 \frac{\omega'^2}{c^2} \beta^2 \right]^{1/2} \tag{54}$$

Dividing Eq. (54) by Eq. (53) and noting that $\mathbf{k}' \cdot \mathbf{v} = k'v \cos \theta'$, we obtain

$$\frac{k}{\omega} = \frac{\sqrt{k'^2 + 2\gamma^2 \frac{\omega'}{c^2} k'v \cos \theta' + (\gamma^2 - 1)k'^2 \cos^2 \theta' + \gamma^2 \frac{\omega'^2}{c^2} \beta^2}}{\gamma(\omega' + k'v \cos \theta')} \tag{55}$$

Since by definition $n = ck/\omega$ and $n' = ck'/\omega'$, it follows from Eq. (55) that

$$n = \frac{ck}{\omega} = \frac{\sqrt{n'^2 + 2\gamma^2 n'\beta \cos \theta' + (\gamma^2 - 1)n'^2 \cos^2 \theta' + \gamma^2\beta^2}}{\gamma(1 + n'\beta \cos \theta')} \tag{56}$$

Although this relation relates n to n', it is not yet the relation we want, because it involves the angle θ'. To obtain the desired relation, we must eliminate θ' in favor of the angle θ between $\mathbf{k}$ and $\mathbf{v}$. Accordingly, we invoke the aberration relations

$$\cos \theta = \frac{\gamma(n' \cos \theta' + \beta)}{\sqrt{n'^2 \sin^2 \theta' + \gamma^2(n' \cos \theta' + \beta)^2}} \tag{57}$$

$$\cos \theta' = \frac{\gamma(n \cos \theta - \beta)}{\sqrt{n^2 \sin^2 \theta + \gamma^2(n \cos \theta - \beta)^2}} \tag{58}$$

which follow from Eq. (46). Combining Eqs. (56) and (57), we are led to

$$n \cos \theta = \frac{n' \cos \theta' + \beta}{1 + n'\beta \cos \theta'} \tag{59}$$

which, with the aid of Eq. (58), yields the following quadratic equation for n:

$$[1 - (n'^2 - 1)\gamma^2\beta^2 \cos^2 \theta]n^2 + [2\gamma^2\beta(n'^2 - 1) \cos \theta]n - \gamma^2(n'^2 - \beta^2) = 0 \tag{60}$$

Solving this equation and choosing the root that yields $n = n'$ for $\mathbf{v} = 0$, we obtain the desired relation:

$$n = \frac{\sqrt{1 + \gamma^2(n'^2 - 1)(1 - \beta^2 \cos^2 \theta)} - \beta\gamma^2(n'^2 - 1) \cos \theta}{1 - \gamma^2(n'^2 - 1)\beta^2 \cos^2 \theta} \tag{61}$$

Here n' is the index of refraction of the medium in the K' frame, which is at rest with respect to the medium, n is the index of refraction in the K frame, with respect to which the medium is moving at velocity $\mathbf{v}$, and θ is the angle between $\mathbf{v}$ and the wave vector $\mathbf{k}$.

We see from Eq. (61) that the index of refraction n of a moving medium depends on the velocity $v(= \beta c)$ of the medium and on the angle θ between $\mathbf{k}$ and $\mathbf{v}$. When $\beta^2 \ll 1$, Eq. (61) reduces to the following equation,

$$n = n' - (n'^2 - 1)\beta \cos \theta \tag{62}$$

which is valid for dispersive as well as nondispersive media. In the case where the direction of $\mathbf{k}$ is parallel ($\theta = 0$) or antiparallel ($\theta = \pi$) to $\mathbf{v}$ and the medium is nondispersive, Eq. (62) yields

$$v_{\mathrm{ph}} = \frac{c}{n'} \pm v\left(1 - \frac{1}{n'^2}\right) \tag{63}$$

where $v_{\mathrm{ph}}(= c/n)$ is the phase velocity of the wave in K. This is the well-known formula of Fresnel. The coefficient $(1 - 1/n'^2)$ is called the Fresnel drag coefficient. The Fresnel formula was verified experimentally by Fizeau who used streaming water as the moving medium.

For a dispersive medium Eq. (63) has to be modified. To find what this modification is, we note that in Eq. (62) the index of refraction n' is a function of ω'. Since the Doppler formula (47) for low velocities ($\beta^2 \ll 1$) yields $\omega' = \omega \mp \beta n\omega$, where the upper sign is for $\theta = 0$ and the lower one is for $\theta = \pi$, we see that

$$n'(\omega') = n'(\omega \mp \beta n\omega) \tag{64}$$

Expanding this relation about ω and keeping only the first two terms, we get

$$n'(\omega') = n'(\omega) \mp \beta n\omega \frac{\partial n'(\omega)}{\partial \omega} \tag{65}$$

Substituting this expansion into the equation

$$n = n' \mp (n'^2 - 1)\beta \tag{66}$$

which follows from Eq. (62) when $\theta = 0$ and $\theta = \pi$, and neglecting terms in β^2, we get

$$n = n'(\omega) \mp [n'^2(\omega) - 1]\beta \mp \beta\omega n'(\omega) \frac{\partial n'(\omega)}{\partial \omega} \tag{67}$$

Since $v_{\mathrm{ph}} = c/n$, we then deduce from Eq. (67) that

$$v_{\mathrm{ph}} = \frac{c}{n'(\omega)} \pm v\left[1 - \frac{1}{n'^2(\omega)}\right] \pm v\frac{\omega}{n'(\omega)}\frac{\partial n'(\omega)}{\partial \omega} \tag{68}$$

This is the form that Eq. (63) takes for a dispersive medium. We see that the dispersive nature of the medium is accounted for by the last term on the right side. This term is sometimes referred to as the "Lorentz term". It was verified experimentally by Zeeman.

7.6 Wave Equation for Moving Homogeneous Isotropic Media

In a frame of reference K' which is at rest with respect to a homogeneous isotropic medium, the vector potential $\mathbf{A}'(\mathbf{r}',t')$ and the scalar potential $\phi'(\mathbf{r}',t')$ due to a current density $\mathbf{J}'(\mathbf{r}',t')$ and a charge density $\rho'(\mathbf{r}',t')$ clearly must obey the inhomogeneous wave equations

$$\left[\nabla'^2 - \frac{n'^2}{c^2}\frac{\partial^2}{\partial t'^2}\right]\mathbf{A}'(\mathbf{r}',t') = -\mu'\mathbf{J}'(\mathbf{r}',t') \tag{69}$$

$$\left[\nabla'^2 - \frac{n'^2}{c^2}\frac{\partial^2}{\partial t'^2}\right]\phi'(\mathbf{r}',t') = -\frac{1}{\epsilon'}\rho'(\mathbf{r}',t') \tag{70}$$

where μ' and ϵ' are the permeability and the dielectric constant of the medium and n' is the index of refraction. With the aid of the 4-vectors J'_α and A'_α, whose values are given by

$$J'_\alpha = \begin{bmatrix} J'_x \\ J'_y \\ J'_z \\ ic\rho' \end{bmatrix} \qquad A'_\alpha = \begin{bmatrix} A'_x \\ A'_y \\ A'_z \\ \frac{i}{c}\phi' \end{bmatrix} \tag{71}$$

these equations can be combined to give

$$(-1 - \kappa c^2\delta_{4\alpha})\left(\nabla'^2 - \frac{n'^2}{c^2}\frac{\partial^2}{\partial t'^2}\right)A'_\alpha = \mu' J'_\alpha \tag{72}$$

where $\kappa = \mu'\epsilon' - (1/c^2) = (n'^2 - 1)/c^2$ and $\delta_{4\alpha}$ is the Kronecker delta.

We wish to transform Eq. (72) to reference frame K, with respect to which the medium is moving at velocity $\mathbf{v}$. Since A'_α and J'_α are 4-vectors, they transform as follows:

$$A'_\alpha = a_{\alpha\beta} A_\beta \qquad J'_\alpha = a_{\alpha\beta} J_\beta \tag{73}$$

Here A_β and J_β are 4-vectors in K, and the $a_{\alpha\beta}$ are the coefficients of the proper Lorentz transformation that carries K' into K. To transform the differential operator that appears in Eq. (72), we write

$$\nabla'^2 - \frac{n'^2}{c^2}\frac{\partial^2}{\partial t'^2} = \nabla'^2 - \frac{1}{c^2}\frac{\partial^2}{\partial t'^2} - \frac{n'^2 - 1}{c^2}\frac{\partial^2}{\partial t'^2} \tag{74}$$

The first two terms on the right side constitute an invariant operator, and hence

$$\nabla'^2 - \frac{1}{c^2}\frac{\partial^2}{\partial t'^2} = \nabla^2 - \frac{1}{c^2}\frac{\partial^2}{\partial t^2} \tag{75}$$

By means of the transformations

$$t = \gamma\left(t' + \frac{\mathbf{r}' \cdot \mathbf{v}}{c^2}\right) \tag{76}$$

$$\mathbf{r} = \mathbf{r}' + \gamma \mathbf{v} t' + (\gamma - 1)\frac{\mathbf{r}' \cdot \mathbf{v}}{v^2}\mathbf{v} \tag{77}$$

it can be shown that

$$\frac{n'^2 - 1}{c^2}\frac{\partial^2}{\partial t'^2} = \kappa\gamma^2\left(\frac{\partial}{\partial t} + \mathbf{v} \cdot \nabla\right)^2 \tag{78}$$

Thus from relations (75) and (78) we see that the operator (74) transforms as follows:

$$\nabla'^2 - \frac{n'^2}{c^2}\frac{\partial^2}{\partial t'^2} = \nabla^2 - \frac{1}{c^2}\frac{\partial^2}{\partial t^2} - \kappa\gamma^2\left(\frac{\partial}{\partial t} + \mathbf{v} \cdot \nabla\right)^2 \tag{79}$$

Now, with the aid of the transformations (73) and (79), it becomes evident that equation (72) in K' transforms into the following equation in K:

$$a_{\alpha\beta}LA_\beta = -\mu' a_{\alpha\beta}J_\beta - \kappa c^2 \delta_{4\alpha} a_{\alpha\beta} LA_\beta \tag{80}$$

where the operator L is defined by

$$L \equiv \nabla^2 - \frac{1}{c^2}\frac{\partial^2}{\partial t^2} - \kappa\gamma^2\left(\frac{\partial}{\partial t} + \mathbf{v}\cdot\nabla\right)^2 \tag{81}$$

Multiplying Eq. (80) by $a_{\alpha\nu}$, summing on α, and using the orthogonality relation (3), we find that

$$LA_\nu = -\mu' J_\nu - \kappa c^2 a_{4\nu} a_{4\beta} LA_\beta \tag{82}$$

For $\alpha = 4$, Eq. (80) yields

$$a_{4\beta}LA_\beta = -\frac{\mu'}{1+\kappa c^2}\, a_{4\beta}J_\beta \tag{83}$$

Therefore we can cast Eq. (82) in the form

$$LA_\nu = -\mu' J_\nu + \frac{\mu'\kappa c^2}{1+\kappa c^2}\, a_{4\nu}a_{4\beta}J_\beta \tag{84}$$

Using Eq. (19), we see that

$$a_{4\nu}a_{4\beta}J_\beta = -\frac{1}{c^2}\, U_\nu J_\beta U_\beta \tag{85}$$

where U_ν is the velocity 4-vector $(\gamma\mathbf{v}, i\gamma c)$. With the aid of this result, Eq. (84) yields

$$LA_\nu = -\mu' J_\nu - \frac{\mu'\kappa}{n'^2}\, U_\nu J_\beta U_\beta \tag{86}$$

This is the equation into which Eq. (72) is transformed when the frame of reference is changed from K' to K.

In three-dimensional form, Eq. (86) leads to the following equations for the vector potential $\mathbf{A}(\mathbf{r},t)$ and the scalar potential $\phi(\mathbf{r},t)$ in reference frame K:

$$\left[\nabla^2 - \frac{1}{c^2}\frac{\partial^2}{\partial t^2} - \kappa\gamma^2\left(\frac{\partial}{\partial t} + \mathbf{v}\cdot\nabla\right)^2\right]\mathbf{A}(\mathbf{r},t)$$
$$= -\mu'\mathbf{J} - \frac{\mu'\kappa}{n'^2}\gamma\mathbf{v}(\gamma\mathbf{J}\cdot\mathbf{v} - \gamma c^2\rho) \quad (87)$$

$$\left[\nabla^2 - \frac{1}{c^2}\frac{\partial^2}{\partial t^2} - \kappa\gamma^2\left(\frac{\partial}{\partial t} + \mathbf{v}\cdot\nabla\right)^2\right]\phi(\mathbf{r},t)$$
$$= -\mu' c^2\rho - \frac{\mu'\kappa}{n'^2}\gamma c^2(\gamma\mathbf{J}\cdot\mathbf{v} - \gamma c^2\rho) \quad (88)$$

where, as before, $\kappa = (c^2\epsilon'\mu' - 1)/c^2 = (n'^2 - 1)/c^2$. With a knowledge of these equations, we can find the vector and scalar potentials of a source surrounded by a homogeneous isotropic medium moving at a velocity $\mathbf{v}$ with respect to the source. Moreover, these equations enable one to calculate the electric vector $\mathbf{E} = -\nabla\phi - (\partial/\partial t)\mathbf{A}$ and the magnetic vector $\mathbf{B} = \nabla \times \mathbf{A}$ of the source in the presence of a wind.

The above discussion is based on the transformation of the inhomogeneous wave equation from the K' frame to the K frame. Actually, the same results can be achieved by using the tensor form of Maxwell's equations as the point of departure.[1] To show this, we recall that Maxwell's equations can be written as follows:

$$\frac{\partial F_{\alpha\beta}}{\partial x_\nu} + \frac{\partial F_{\beta\nu}}{\partial x_\alpha} + \frac{\partial F_{\nu\alpha}}{\partial x_\beta} = 0 \quad (89)$$

$$\frac{\partial G_{\alpha\beta}}{\partial x_\beta} = J_\alpha \quad (90)$$

These tensor equations hold in all Lorentz frames, and in particular they hold in K' and K. In K' the constitutive relations are

$$\mathbf{D}' = \epsilon'\mathbf{E}' \qquad \text{and} \qquad \mathbf{H}' = \frac{1}{\mu'}\mathbf{B}' \quad (91)$$

[1] K. S. H. Lee, On the Doppler Effect in a Medium, *Antenna Lab. Rept.* 29, California Institute of Technology, December, 1963. See also, J. M. Jauch and K. M. Watson, Phenomenological Quantum-electrodynamics, *Phys. Rev.*, **74**: 950, 1485 (1948).

Expressing **D**′, **E**′, **H**′, **B**′ in terms of **D**, **E**, **H**, **B** of the reference frame K, we find with the aid of the Eqs. (24), (25), (26), and (27) that the constitutive relations in K are

$$\mathbf{D} + \frac{1}{c^2}\mathbf{v} \times \mathbf{H} = \epsilon'(\mathbf{E} + \mathbf{v} \times \mathbf{B}) \tag{92}$$

$$\mathbf{H} - \mathbf{v} \times \mathbf{D} = \frac{1}{\mu'}\left(\mathbf{B} - \frac{1}{c^2}\mathbf{v} \times \mathbf{E}\right) \tag{93}$$

When written in tensor form, these constitutive relations become

$$G_{\alpha\beta}U_\beta = c^2\epsilon' F_{\alpha\nu}U_\nu \tag{94}$$

$$G_{\alpha\beta}U_\nu + G_{\beta\nu}U_\alpha + G_{\nu\alpha}U_\beta = \frac{1}{\mu'}(F_{\alpha\beta}U_\nu + F_{\beta\nu}U_\alpha + F_{\nu\alpha}U_\beta) \tag{95}$$

where as before U_ν denotes the velocity 4-vector $(\gamma\mathbf{v}, i\gamma c)$.

To express the field tensor $G_{\alpha\beta}$ explicitly in terms of the field tensor $F_{\alpha\beta}$, we multiply Eq. (95) by U_ν. Noting that

$$U_\nu U_\nu = -c^2$$

we thus find that

$$-c^2G_{\alpha\beta} + U_\alpha G_{\beta\nu}U_\nu + U_\beta G_{\nu\alpha}U_\nu = \frac{1}{\mu'}\{-c^2F_{\alpha\beta} + U_\alpha F_{\beta\nu}U_\nu + U_\beta F_{\nu\alpha}U_\nu\} \tag{96}$$

By virtue of the constitutive relations (94), we have

$$G_{\beta\nu}U_\nu = c^2\epsilon' F_{\beta\nu}U_\nu \qquad G_{\nu\alpha}U_\nu = -c^2\epsilon' F_{\alpha\nu}U_\nu \tag{97}$$

Hence, it follows from Eq. (96) that

$$G_{\alpha\beta} = \frac{1}{\mu'}F_{\alpha\beta} + \frac{\kappa}{\mu'}(U_\alpha F_{\beta\nu}U_\nu - U_\beta F_{\alpha\nu}U_\nu) \tag{98}$$

Substituting expression (98) into the Maxwell equation (90), we find that

$$\frac{\partial F_{\alpha\beta}}{\partial x_\beta} + \kappa U_\alpha U_\nu \frac{\partial F_{\beta\nu}}{\partial x_\beta} - \kappa U_\nu U_\beta \frac{\partial F_{\alpha\nu}}{\partial x_\beta} = \mu' J_\alpha \tag{99}$$

However, from Eqs. (90) and (94) we see that

$$\kappa U_\alpha U_\nu \frac{\partial F_{\beta\nu}}{\partial x_\beta} = -\frac{\kappa}{c^2\epsilon'} U_\alpha J_\beta U_\beta \tag{100}$$

Therefore, Eq. (99) becomes

$$\frac{\partial F_{\alpha\beta}}{\partial x_\beta} - \kappa U_\nu U_\beta \frac{\partial F_{\alpha\nu}}{\partial x_\beta} = \mu' J_\alpha + \frac{\kappa}{c^2\epsilon'} U_\alpha J_\beta U_\beta \tag{101}$$

Now we have two equations for the field tensor $F_{\alpha\beta}$, one being the Maxwell equation (89) and the other being equation (101). If we write the field tensor $F_{\alpha\beta}$ in terms of the 4-potential $A_\nu = \left(\mathbf{A}, i, \frac{\phi}{c}\right)$, that is, if we write

$$F_{\alpha\beta} = \frac{\partial A_\beta}{\partial x_\alpha} - \frac{\partial A_\alpha}{\partial x_\beta} \tag{102}$$

then Eq. (89) is satisfied. Substituting expression (102) into Eq. (101), we obtain the following equation for the 4-potential:

$$\frac{\partial}{\partial x_\beta}\frac{\partial}{\partial x_\alpha} A_\beta - \frac{\partial}{\partial x_\beta}\frac{\partial}{\partial x_\beta} A_\alpha - \kappa U_\nu U_\beta \left(\frac{\partial}{\partial x_\beta}\frac{\partial}{\partial x_\alpha} A_\nu - \frac{\partial}{\partial x_\beta}\frac{\partial}{\partial x_\nu} A_\alpha\right) = \mu' J_\alpha + \frac{\kappa}{c^2\epsilon'} U_\alpha J_\beta U_\beta \tag{103}$$

Rearranging terms, we get

$$\frac{\partial}{\partial x_\alpha}\left(\frac{\partial A_\beta}{\partial x_\beta} - \kappa U_\nu U_\beta \frac{\partial A_\nu}{\partial x_\beta}\right) - \left(\frac{\partial}{\partial x_\beta}\frac{\partial}{\partial x_\beta} - \kappa U_\nu U_\beta \frac{\partial}{\partial x_\beta}\frac{\partial}{\partial x_\nu}\right) A_\alpha = \mu' J_\alpha + \frac{\kappa}{c^2\epsilon'} U_\alpha J_\beta U_\beta \tag{104}$$

Since the 4-potential is not completely determined by Eq. (102), we are free to impose on it the following additional condition,

$$\frac{\partial A_\beta}{\partial x_\beta} - \kappa U_\nu U_\beta \frac{\partial A_\nu}{\partial x_\beta} = 0 \tag{105}$$

which is called the "generalized Lorentz condition" for the 4-potential. When this condition is satisfied, Eq. (104) reduces to

$$\left(\frac{\partial}{\partial x_\beta}\frac{\partial}{\partial x_\beta} - \kappa U_\nu U_\beta \frac{\partial}{\partial x_\beta}\frac{\partial}{\partial x_\nu}\right) A_\alpha = -\mu' J_\alpha - \frac{\kappa}{c^2\epsilon'} U_\alpha J_\beta U_\beta \tag{106}$$

This equation is identical to Eq. (86) and, in three-dimensional form, amounts to Eqs. (87) and (88).

To show that Eq. (106) can be used to find the index of refraction of a moving media, we assume that $\mathbf{A}(\mathbf{r},t)$ has the form of a plane wave:

$$\mathbf{A}(\mathbf{r},t) = \mathrm{Re}\,\{\mathbf{A}_0 e^{i\mathbf{k}\cdot\mathbf{r}} e^{-i\omega t}\} \tag{107}$$

Substituting this expression into Eq. (87), with the right side set equal to zero, we find

$$-c^2k^2 + \omega^2 + \gamma^2(n'^2 - 1)(\mathbf{k}\cdot\mathbf{v} - \omega)^2 = 0 \tag{108}$$

On solving this equation for $n = ck/\omega$, we are led to relation (61) for the index of refraction.

Name Index

Subject Index

A CATALOG OF SELECTED

DOVER BOOKS

IN SCIENCE AND MATHEMATICS

A CATALOG OF SELECTED
DOVER BOOKS
IN SCIENCE AND MATHEMATICS